Influence of Internal States on Ion-Molecule Reactions in a Temperature Variable 22-Pole Ion Trap

-

Spectroscopy and Reaction Kinetics

Influence of Internal States on Ion-Molecule Reactions in a Temperature Variable 22-Pole Ion Trap

-

Spectroscopy and Reaction Kinetics

Inaugural-Dissertation

zur
Erlangung des Doktorgrades
der Mathematisch-Naturwissenschaftlichen Fakultät
der Universität zu Köln

vorgelegt von

Sabrina Gärtner

aus Leverkusen

Cuvillier Verlag, Göttingen

2014

Bibliografische Information der Deutschen Nationalbibliothek

Die Deutsche Nationalbibliothek verzeichnet diese Publikation in der Deutschen Nationalbibliographie; detaillierte bibliographische Daten sind im Internet über http://dnb.d-nb.de abrufbar.

1. Aufl. - Göttingen: Cuvillier, 2014

Zugl.: Köln, Univ., Diss., 2014

Berichterstatter:	Prof. Dr. Stephan Schlemmer
(Gutachter)	apl. Prof. Dr. Andreas Wolf

Tag der mündlichen Prüfung: 02.12.2013

© CUVILLIER VERLAG, Göttingen 2014
Nonnenstieg 8, 37075 Göttingen
Telefon: 0551-54724-0
Telefax: 0551-54724-21
www.cuvillier.de

ISBN 978-3-95404-689-8

eISBN 978-3-7369-4689-7

Contents

List of Figures

List of Tables

Chapter 1

Kurzzusammenfassung

Im interstellaren Medium sind chemische Reaktionen zwischen Ionen und neutralem Gas von großer Bedeutung. Da sie üblicherweise deutlich geringere Barrieren besitzen als Reaktionen zwischen zwei neutralen Partnern, sind sie bei niedrigen Temperaturen bevorzugt.

Die vorliegende Arbeit befasst sich mit Untersuchungen an den endothermen Reaktionen von CH_2D^+, H_2D^+ und N^+ mit H_2. Sie werden bei tiefen Temperaturen stark von der Translationsenergie und von der internen Anregung der Reaktionspartner beeinflusst. Diese empfindliche Abhängigkeit kann beispielsweise dazu verwendet werden, Informationen über Lage oder Besetzung der Energieniveaus der Reaktanden zu erhalten.

Die hier vorgestellten Untersuchungen wurden mit Hilfe eines 22-Pol Ionenspeichers durchgeführt, der bei Temperaturen zwischen 10 und 300 K betrieben werden kann. Im Speicher können Drücke eingestellt werden, die niedrig genug sind, um chemische Prozesse unter typischen astrophysikalischen Bedingungen zu untersuchen. Der Aufbau dieser Apparatur, sowie drei verschiedene Arten von Messungen, die mit ihr durchgeführt werden können (Massenspektroskopie, Reaktionskinetik, Spektroskopie an Ionen), werden in Kapitel 4 erläutert. Außerdem enthält dieses Kapitel eine kurze Beschreibung der Konstruktion von neuen Daly Detektoren für die beiden neuen Ionenspeicher-Experimente in Köln.

Alle endothermen Reaktionen mit Wasserstoff werden bei tiefen Temperaturen unter anderem von der Kernspinkonfiguration von H_2 beeinflusst. Daher ist es für solche Untersuchungen von großer Bedeutung, die Zusammensetzung der Wasserstoffprobe aus den beiden möglichen Konfigurationssymmetrien *ortho* und *para* zu kennen und diese gezielt beeinflussen zu können.

Kapitel 5 beschreibt die durchgeführten Untersuchungen des ortho-zu-para Verhältnisses und Verbesserungen am Kölner Para-Wasserstoff-Generator, die im Rahmen dieser Arbeit vorgenommen wurden. Um das ortho-zu-para Verhältnis zu bestimmen, wurden zwei verschiedene Methoden angewandt, zum Einen Raman Spektroskopie, zum Anderen der Einfluss der Rotationsenergie von ortho-Wasserstoff auf chemische Gleichgewichte bei tiefen Temperaturen ($H_2D^+ + H_2$, $N^+ + H_2$). Die Ergebnisse der Raman Spektroskopie und des Gleichgewichts der Reaktion von H_2D^+ mit H_2 werden ebenfalls in Kapitel 5 vorgestellt. Die Ergebnisse der Reaktion von N^+ mit H_2 sind in Kapitel 6 beschrieben, da dieses Reaktionssystem detaillierter untersucht wurde.

Die Reaktion von N^+ mit H_2 ist der erste Schritt zur Bildung von interstellarem Ammoniak. Sie wird seit drei Jahrzehnten mit verschiedenen Methoden von einer Reihe von Gruppen untersucht. Dennoch sind nach wie vor viele Fragen offen. So ist beispielsweise ungeklärt, ob die Reaktion tatsächlich endotherm ist, oder ob sie lediglich von einer Barriere behindert wird. Auch ist nicht eindeutig geklärt, in welchem Maße die Feinstruktur-Energie von N^+ zur Unterstützung der Reaktion verfügbar ist. Messungen zu diesem Reaktionssystem finden sich in Kapitel 6.

Zu Beginn dieser Arbeit stand die mutmaßliche Entdeckung von CH_2D^+ im All. Die bisher verfügbaren Vorhersagen für seine Rotationsübergänge waren nicht präzise genug, um die Entdeckung zu bestätigen. Da die Methode der Laserinduzierten Reaktionen (LIR) hochaufgelöste Spektren von Ionen liefern kann, wurde CH_2D^+ im Ionenspeicherexperiment LIRTrap spektroskopisch untersucht. In Kapitel 7 werden die durchgeführten Infrarot-Messungen an zwei Vibrationsbanden von CH_2D^+ und die resultierenden Vorhersagen für reine Rotationsübergänge vorgestellt.

Häufig beobachtet man bei LIR-Linien Abweichungen von der Gauss'schen Linienform. Solche Abweichungen behindern glücklicherweise nicht die Frequenzbestimmung, da die mathematische Beschreibung der beobachteten Linienformen qualitativ bekannt ist. Der Zusammenhang zwischen den Parametern der Funktion, die solche gesättigten Linien beschreibt, und denen des Experiments ist jedoch unbekannt. Daher ist es noch nicht möglich, weitere Informationen wie Translations- oder Rotationstemperatur der Ionen aus den gesättigten Linien zu bestimmen. Kapitel 8 befasst sich mit der quantitativen Beschreibung des LIR Prozesses und stellt numerische Simulationen sowie Messungen an CH_5^+ vor, die zu diesem Thema durchgeführt wurden.

Im Rahmen dieser Arbeit wurde die erwähnte Empfindlichkeit endothermer Reaktionen gegenüber internen Anregungen der Reaktionspartner für zwei verschiedene Untersuchungtypen genutzt. Sie ermöglichte hochauflösende Infrarotspektroskopie an CH_2D^+ und damit verbesserte Vorhersagen für reine Rotationsübergänge dieses Ions. Erste Schritte zum detaillierten Verständnis der Prozesse bei laserinduzierten Reaktionen wurden gemacht. Außerdem ermöglichte diese Empfindlichkeit, die Bestimmung der Qualität von selbst hergestellten para-Wasserstoff-Proben und Einblicke in die Reaktionskinetik des Systems $N^+ + H_2$.

Chapter 2

Abstract

Chemical reactions between ions and neutral gas are very important in the interstellar medium. This is due to the fact, that they usually have much lower barriers than neutral-neutral reactions. Thus, they are more likely to take place at low temperatures.

In this work, the endothermic reactions of CH_2D^+, H_2D^+, and N^+ with H_2 have been investigated. At low temperatures, they depend very sensitively on the translational and internal energies of the reaction partners. This sensitivity can be applied to gain information on the energy levels or the population of the internal states.

The investigations were performed in a temperature variable (10 to 300 K) 22-pole ion trap. Inside the trap, the pressure is low enough, to investigate chemical processes under typical astrophysical conditions. The setup of the ion trap apparatus and three different types of measurements that can be performed in it (mass spectroscopy, reaction kinetics, spectroscopy of ions) are presented in Chapter 4. This chapter also includes the brief description of the assembly of new Daly detectors for the two new ion trap experiments in Cologne.

For the investigation of all endothermic reactions of ions with molecular hydrogen at low temperatures, it is important to know the fractions of each of the two symmetries of nuclear spin configuration *ortho* and *para* in H_2.

Investigations on the ortho-to-para ratio and improvements on the Cologne para hydrogen generator performed in this work are described in Chapter 5. Two different methods can be applied in Cologne to test this ratio, Raman spectroscopy and the influence of the different internal energies of ortho and para hydrogen on chemical equilibria at low temperatures ($H_2D^+ + H_2$, $N^+ + H_2$). The results of Raman spectroscopy and of the equilibrium of the reaction of H_2D^+ with H_2 are also presented in Chapter 5. The results from the reaction of

N^+ with H_2 are presented in Chapter 6 as this reaction system was investigated in more detail.

The reaction of N^+ with H_2 is the first step in the formation of interstellar ammonia. It has been investigated with various methods by several groups over the last three decades, but still many open questions remain. For example, it is not known whether the reaction is really endothermic or just hindered by a barrier, and to what extend the energy of the fine-structure states of N^+ is available to support the reaction. Measurements on this reaction system are presented in Chapter 6.

The recent tentative astronomical detection of CH_2D^+ called for more precise predictions of the ion's pure rotational lines. Thus, the spectroscopy of CH_2D^+ was revived in several groups. Measurements on the rovibrational transitions of two vibrational bands were performed in the ion trap, as high resolution spectra can be obtained by the method of laser induced reactions (LIR). These measurements and the resulting predictions for pure rotational transitions are presented in Chapter 7.

Many lines observed with LIR show deviations from the Gaussian line profile. These deviations do not hinder the precise frequency determination, as their mathematical description is known qualitatively. As the correlation between the experimental parameters and those of the function for the saturated LIR lines is unknown, it is not possible to derive other information such as translational and rotational temperatures of the ions from the observed non-Gaussian lines. Investigations towards a quantitative description of the involved processes are presented in Chapter 8.

In this work, the already mentioned sensitivity of endothermic reactions on internal excitations of the reaction partners was applied for two different types of investigations. It enables high resolution spectroscopy of CH_2D^+ and thereby improved predictions for pure rotational transitions of this ion. First steps were made towards a detailed understanding of the processes in laser induced reactions. Additionally, the explained sensitivity allowed to determine the purity of para hydrogen samples and yielded insight into the reaction kinetics of the system $N^+ + H_2$.

Chapter 3

Introduction

Light is the only information we have to understand processes like star formation and the origin of the chemical compositions of earth and other objects in our own solar system up to the origin of life. From the light we observe, we have to deduce everything. Thus, the first step of course is, to identify the spectral features in astronomical observations and assign them to matter (atoms, molecules, ions, grains). The second step is to derive physical properties of the emitting or absorbing particles, e.g. their temperature, their abundance, non-thermal velocities, etc.. The third step is to learn about the history of the observed region from the present information. Here, knowledge of the astrochemical processes can be very helpful. If all reaction mechanisms and rates were known, one could tell from the current composition of an observed object, what chemical evolution it probably ran through. This would allow to determine for example the age of the object.

In the laboratory astrophysics group in Cologne, spectroscopy as well as astrochemistry are pursued.

Astrochemistry is very different from chemistry on earth. The interstellar medium is usually very thin and cold compared to the typical environment on earth. The low temperatures hinder all reactions that are strongly endothermic or have high energy barriers. Thus, exothermic reactions with low barriers can be expected to have the highest rates at temperatures of some ten Kelvin. Usually, the barriers for reactions between an ion and a neutral molecule are much lower than those for the reaction between two neutrals. As ions are easily produced in space by gamma rays, free electrons, etc., ion-neutral reactions play a much stronger role in space, than they do on earth.

Another peculiarity of astrochemistry is the low number density. Many exothermic reactions have much lower rates than one would expect, because in two-body-collisions, there is no chance to get rid of the released energy. Three-body-collisions increase the rate of such reactions dramatically (the third collision partner carries away the released energy), but they can be excluded for most interstellar regions due to the low number densities.

A temperature variable ion trap apparatus (LIRTrap) is available in Cologne, two more have been built since this work started. LIRTrap allows to investigate ion-molecule reactions at temperatures between 10 and 300 K and to obtain high resolution spectra of ions.

In this work we describe several different experiments performed in the ion trap. All of them apply the very sensitive dependence of endothermic reactions on translational and internal energies of the reaction partners at low temperatures to obtain information about the reaction partners. This sensitivity allows for example to perform spectroscopy of cold trapped ions by shifting the reaction equilibrium via laser irradiation. The method is called laser induced reactions (LIR) and has been performed on several species in the past. In this work, two vibrational bands of CH_2D^+ have been investigated.

Additionally, efforts were made to derive a quantitative description of the LIR process itself. LIR lines often show deviations from a Gaussian line profile. Although a function is at hand to describe these profiles, it cannot be connected to the experimental parameters (with the frequency being the only exclusion). A full description of the LIR process will probably include so called state specific rate coefficients. Those constants tell with which rate two specific levels of the involved ion and neutral will react to form the products.

State specific rate coefficients have long been searched for, for several basic reactions of astrochemistry. They would allow precise predictions of chemical evolutions even of regions that are far away from thermal equilibrium, as long as the level population of the involved species is known (e.g. by spectroscopic observations). An important step in the quest for state specific rate coefficients is to reduce the number of involved states. Here, molecular hydrogen has a useful feature. It can exist in two different nuclear spin configurations. For symmetry reasons, they have to occupy different rotational levels. Therefore, a para hydrogen sample that is cooled down to ten Kelvin will occupy basically just the rotational ground state. As hydrogen comes at room temperature in a mixture of ortho and para, and spin flips are strictly forbidden, some efforts have to be made to produce para hydrogen.

The purity of the produced para hydrogen sample can be investigated in different ways. The already mentioned dependence of endothermic reactions on internal energies is one of them, Raman spectroscopy another. Both methods were used in this work. Two possible test reactions are $H_2D^+ + H_2 \rightleftharpoons H_3^+ + HD$ and $N^+ + H_2 \rightleftharpoons NH^+ + H$. Both have been applied in this work.

The latter reaction was investigated in great detail. It is the first step in the formation of interstellar ammonia and still provides challenges for laboratory astrophysics even after three decades of investigations.

Chapter 4

LIRTrap

Reactions between ions and neutral molecules are very important for astrochemistry. This is due to the fact, that they usually have much lower barriers than reactions between two neutral partners. Therefore, they are much more likely to happen at low temperatures. The 22-pole ion trap apparatus that will be described in the following section enables us to investigate the kinetics of such ion-molecule reactions at temperatures down to ≈ 10 K and gas densities between 10^{10} and 10^{15} cm^{-3}. The chemistry under such conditions is comparable to that in e.g. cold dense molecular clouds (10 to 50 K, 10^4 to 10^7 cm^{-3}, the different types of interstellar regions are summarized e.g. in the introduction of [1]). Although the number density of the reaction gas is some orders of magnitude higher in the ion trap, it can be chosen low enough ($\approx 10^{11}$ cm^{-3}) to exclude three body collisions. Thus, the reactions are accelerated compared to molecular clouds, but the reaction mechanisms stay the same. The low temperatures and number densities in the trap also allow to obtain high resolution spectra of ions of astrophysical relevance, such as H_2D^+ and CH_2D^+.

4.1   Setup of the 22-Pole Trap

The setup of the 22-pole ion trap apparatus is sketched in Fig. 4.1. The main parts are the ion source, two quadrupole mass filters, the 22-pole trap itself and the Daly detector. Detailed descriptions of the parts can be found in [4, 5] and references therein. The current version of the 22-pole trap is described in [6].

In the ion storage source, permanently various ions are produced by electron impact on a neutral precursor gas. They are stored in the source until the next experimental cycle. At the beginning of each experimental cycle, a pulse of ions is extracted from the source into the first quadrupole mass filter. The time between two extractions of ions will be called *cycle time* in the following.

The quadrupole selects the desired mass to charge ratios for the experiment from the variety of ions produced in the source. Since in the 22-pole trap experi-

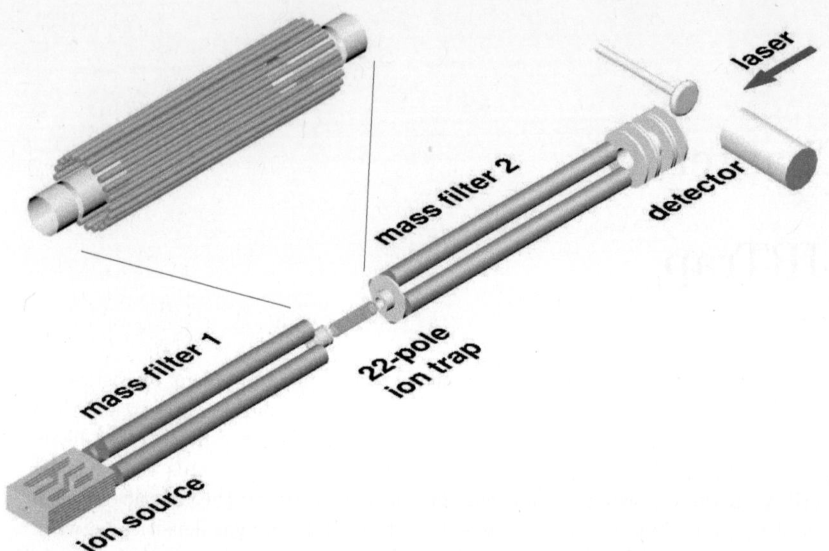

Figure 4.1: Schematic of the 22-pole ion trap (from [2]). From left to right:
Ion Storage Source: Production of ions by electron impact on a neutral gas.
Quadrupole Mass Filter 1: Selection of the mass range to be guided to the trap.
Einzel Lens: Focusing and directing lenses guide the ion cloud into the 22-pole.
22-Pole Trap: The ion cloud is stored together with a neutral reaction gas to perform experiments. Electrodes at both ends confine the axial motion of the ions (see also enlarged view). Storage times are typically several hundred ms. The trap is mounted on a cold head. At the beginning of the storage time, a dense pulse of cold He is emitted by a Piezo valve. The helium cloud cools down the ions to the cryogenic temperatures.
Lens: Focusing and directing lenses guide the ion cloud into the second quadrupole mass filter.
Quadrupole Mass Filter 2: Selects exactly one mass from the charged reaction products, to be counted in the detector.
Acceleration Lenses: Accelerate the ions towards the detector.
Daly Detector [3]: The ions are accelerated towards an aluminium knob by a voltage of -35 kV. There they produce electrons on impact, which are accelerated onto a scintillator, where they produce photons. These photons are then counted in a photomultiplier tube.
Laser: The whole setup is axially transparent, allowing spectroscopy with laser beams entering from the source and/or the detector side of the trap apparatus.

ments, the ions always carry exactly one elementary charge, it is sufficient to talk about mass selection in the following. There are two basic operation modes for the mass selection in quadrupoles. One is the high pass mode, in which all ions above a certain mass are guided through the quadrupole, while lighter ones are discarded. The second one is the selection of a certain mass range. The width of this range can be several or just one atomic mass unit. The ions of the selected mass(es) are guided to the ion trap. At the end of the first quadrupole, an Einzel lens focuses and directs the ion cloud into the 22-pole trap.

Inside the trap, the ion cloud is stored together with a neutral reaction gas to perform experiments. As the 22-pole confines the motion of the ions only in radial direction, two more electrodes are needed to reflect the ions at either end of the 22-pole. At the beginning and at the end of the storage time, the respective potential is decreased to let the ions enter/leave the trap. Storage times are typically several hundred ms. The trap is mounted on a closed cycle helium refrigerator to allow experiments at cryogenic temperatures (down to nominal temperatures of 10 K). For temperature control, a heating wire[1] is attached to the cold head, allowing stable nominal temperatures in the range of 10 to 30 K. Higher temperatures can be reached by switching on and off the cold head at intervals.

At the beginning of the storage time, usually a dense pulse of cold helium is present in the trap (controlled by a Piezo valve), to quickly cool down the ions to the ambient cryogenic temperatures. The neutral reaction gas is usually constantly admitted to the trap. The reaction gas is then continuously flowing through the trap chamber. Some experiments require modifications to that scheme. When the rate for thermalization processes has to be taken into account in the evaluation of measurement data, the helium buffer gas needs to be admitted to the ion trap in continuous flow instead of a pulse, so that its number density is known. When the reaction gas is freezing at the experimental temperatures, it is preferably applied via a second pulsed valve instead of continuous flow, to reduce the freezing rate. The pressures of source (precursor) and trap (reaction and buffer) gases are monitored during the measurements by ionization gauges mounted in the vacuum chambers of the first quadrupole[2] and of the ion trap[3]. As the 22-pole is surrounded by a small chamber and a cold shield, the number density of the reaction gas inside the 22-pole chamber is higher than in the surrounding vacuum chamber (see [7] for more details). The actual number density $[H_2]$ in the innermost chamber can be calculated from the pressure in the trap

---

[1] Kapton heating foil
[2] Arun Microelectronics LTD, Pressure Gauge Controller NGC2 UHV
[3] Arun Microelectronics LTD, Pressure Gauge Controller PGC2 UHV

vacuum chamber p_{trap} and the temperature of the neutral reaction gas (which is approximately the nominal trap temperature T_{trap}) as

$$[\mathrm{H_2}] = 4.18 \times 10^{17} \frac{\sqrt{\mathrm{K}}}{\mathrm{mbar} \cdot \mathrm{cm^3}} \cdot f_{\mathrm{cal}} \cdot \frac{p_{\mathrm{trap}}}{\sqrt{T_{\mathrm{trap}}}} . \tag{4.1}$$

The factor f_{cal} is derived from calibration measurements with a spinning rotor gauge[4] that is mounted to the 22-pole chamber itself.

After the exit of the 22-pole trap, the ions again pass a lens, to be guided into the second quadrupole mass filter[5]. The second quadrupole mass filter selects exactly one mass from the charged reaction products leaving the trap, to be counted in the detector. After leaving the mass filter, the ions are accelerated towards the detector by the acceleration lenses. Then they are counted in the Daly detector. More details on the operation principle of a Daly detector are given in Section 4.5 and can also be found in the original publication of Daly [3].

As is indicated by the red arrow in Figure 4.1, the setup of the 22-pole trap is axially transparent. This allows spectroscopy with laser beams entering from the front (ion source) and/or the back (detector) end of the trap.

The following sections describe the three most important types of measurements, performed in the 22-pole ion trap experiment (mass spectroscopy, reaction kinetics, and spectroscopy of ions) and the assembly of the Daly detectors for the two new ion trap experiments built in Cologne.

---

[4]MKS Instruments, Spinning Rotor Gauge SRG-2CE
[5]Balzers Quadrupol-Massenspektrometer QMG 511

4.2 Mass Spectra

The 22-pole ion trap apparatus allows to take mass spectra of the contents of either the ion source without any storage in the trap or of the ion trap after some storage time. For this measurements, cycle time and storage time are fixed, while for each experimental cycle, the second quadrupole is set to another mass. The width of these mass steps, the mass range to be covered and the number

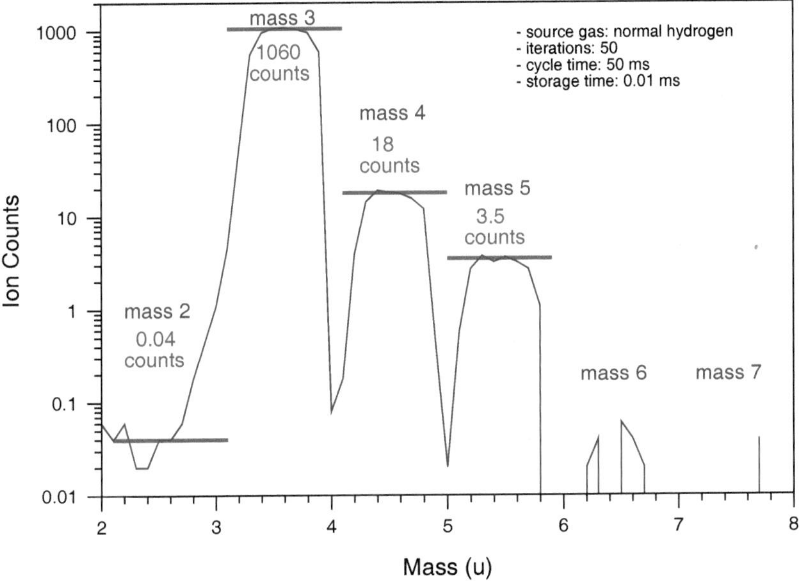

Figure 4.2: Example of a mass spectrum. Normal hydrogen was ionized in the ion source and the produced ions passed through the ion trap without storage. The first quadrupole was not in any selective mode, it served just as a guide for these measurements. The second quadrupole selected the desired masses.
The heights of the different mass peaks give information about the amount of HD the ionized gas contains. The more HD is available in the source, the higher the peaks at masses 4 u (H_2D^+), 5 u (H_5^+, HD_2^+), etc. will be with respect to mass 3 u (H_3^+). The peaks of the masses are somewhat shifted in comparison to the labels on the axis. The mass values are derived from the operation voltages of the second quadrupole. This correlation underlies slow shifts with operation time (over the range of weeks) and is not proportional over the whole mass range of interest (max. 1 - 50 u). Thus, some knowledge of the typical relative intensities of the expected mass peaks is required to assign them properly.

of iterations (how often the chosen mass range is to be scanned) are variable parameters.

Figure 4.2 shows an example of this measurement type. Here, normal hydrogen was admitted to the ion source where it was ionized. The number density in the source was around 3×10^{11} cm^{-3}. The mass range of 2 to 8 u was scanned 50 times in steps of 0.1 u. For these measurements the shortest possible cycle time of 50 ms and no storage time were chosen (the exit of the ion trap opened 0.01 ms after the entrance to the trap opened, meaning that both were open simultaneously for ≈ 250 μs and the ions could pass through the trap directly).

This method allows to investigate what types of ions are produced in the ion source. With no gas admitted to the source, it can be used to search for possible impurities in the source. These impurities might have remained there from previous experiments (like hydrocarbons) or they might leak into the source from the laboratory air (like H_2O (which is always present in very small amounts), N_2, O_2, CO_2). When no precursor gas is available for ionization, all ionization products stem from these impurities, allowing their identification. It can also be applied to investigate the contents of the source gas, as was done in the measurements shown in Figure 4.2. These measurements were performed to compare the HD contents of normal and para hydrogen. As described in Chapter 5.1, hydrogen might loose some of its HD contents in the cold converter box, when it is converted to para hydrogen in the para hydrogen generator. Therefore, normal hydrogen and para hydrogen produced in different operation modes of the para generator were ionized, to look for changes in the ratios of the different isotopologues of H_3^+ ions produced in the ion source. These measurements are presented in Chapter 5.2.

Mass spectra are also very useful, when you need a quick overview of a reaction system. For this purpose, a reaction gas is admitted to the ion trap and the ions are stored in the trap for some time. Taking a broad range mass spectrum at this conditions allows to quickly identify all possible reaction products (also the ones of competing reactions with impurities in the trap). The knowledge of the products that have significant numbers after certain storage times is important to properly choose the experimental settings for the investigation of the temporal evolution of a reaction system. This measurement type will be explained in the next section.

4.3　Time Evolution of Chemical Reactions

The investigation of reaction kinetics for reaction systems of astrochemical relevance is one of the two most important applications of the 22-pole ion trap apparatus [8–10] and will be explained in the following. The other is spectroscopy of ions, which will be explained in Section 4.4.

The temporal evolution of a reaction system can be investigated by varying the storage time in the trap. For this purpose, usually exactly one atomic mass is selected in the first quadrupole mass filter. These ions are then stored together with the neutral reaction gas in the ion trap, which is kept at the temperature of interest. In order to measure the time dependence of the reactions happening

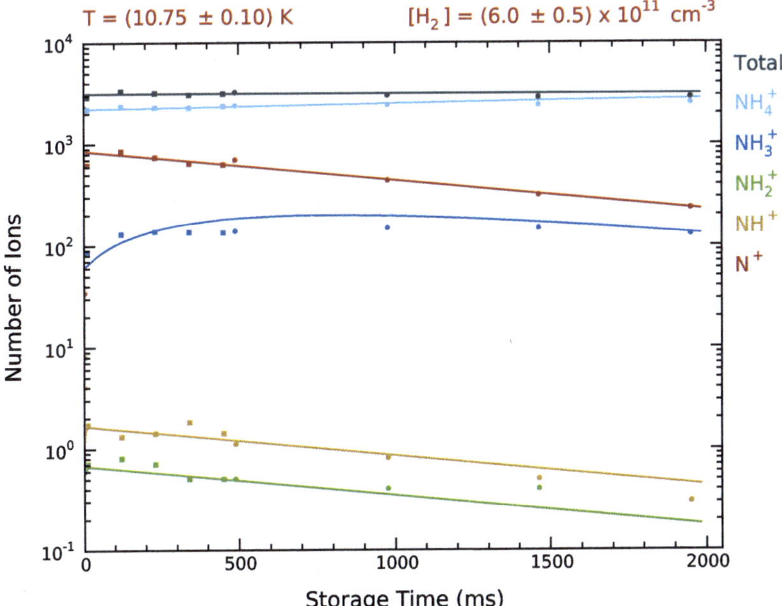

Figure 4.3: Time evolution of the reaction system N^+ + p-H_2 at a nominal temperature of 10.75 K. In each reaction step, an H atom is added to the ion, forming finally NH_4^+ (reactions (4.2) to (4.5)). The solid lines are numerical solutions of the rate equations. The following rate coefficients and ion numbers at $t = 0$ were used for the numerical simulation:
$N^+{}_0 = 850$, $k_{f1} = 1.1 \times 10^{-12}\ \frac{cm^3}{s}$, $NH^+{}_0 = 1$, $k_{f2} = 5.7 \times 10^{-10}\ \frac{cm^3}{s}$, $NH_2^+{}_0 = 1$, $k_{f3} = 1.4 \times 10^{-9}\ \frac{cm^3}{s}$, $NH_3^+{}_0 = 60$, $k_{f4} = 2.8 \times 10^{-12}\ \frac{cm^3}{s}$, $NH_4^+{}_0 = 2200$.

in the ion trap, all expected product masses have to be measured after several storage times. Therefore, the experimental cycle has to be repeated several times with different settings for the second quadrupole mass filter and different storage times. This whole series of measurements is then repeated usually 10 times to improve statistics.

Figure 4.3 shows an example of such a time scan. In this case, N^+ ions were stored in the ion trap together with para hydrogen at a nominal temperature of ≈ 10 K. The following reactions are expected to occur:

$$N^+ + H_2 \underset{k_{b1}}{\overset{k_{f1}}{\rightleftharpoons}} NH^+ + H \tag{4.2}$$

$$NH^+ + H_2 \underset{k_{b2}}{\overset{k_{f2}}{\rightleftharpoons}} NH_2^+ + H \tag{4.3}$$

$$NH_2^+ + H_2 \underset{k_{b3}}{\overset{k_{f3}}{\rightleftharpoons}} NH_3^+ + H \tag{4.4}$$

$$NH_3^+ + H_2 \underset{k_{b4}}{\overset{k_{f4}}{\rightleftharpoons}} NH_4^+ + H \, , \tag{4.5}$$

since no H atoms are present in the trap, we expect only the forward reactions (the reaction gas is flowing through the trap chamber continuously, meaning that all H atoms will be pumped away and be replaced by H_2 molecules immediately; the ions are not pumped out, since they are trapped by the electrical fields). The endothermic reaction (4.2) can be used as a test for the purity of para hydrogen (see Chapter 6). Reactions (4.2) to (4.5) are steps in the formation of interstellar ammonia NH_3.

The number densities of the neutral reaction gas are typically low enough to exclude three body collisions ($\approx 10^{11}$ cm^{-3}). This and the cryogenic temperatures allow to investigate ion-neutral reactions at conditions comparable to those in the interstellar medium (e.g. molecular clouds), but on much shorter timescales (seconds instead of years).

4.4   Spectroscopy via Laser Induced Reactions

As explained in the previous sections, the setup of the 22-pole ion trap apparatus is axially transparent to enable spectroscopy of ions. Since the number of trapped ions is usually on the order of 1000, classic absorption spectroscopy would not be possible. Therefore, action spectroscopy is applied. This means, that in order to observe the absorption of a photon not the number of photons is detected, but that changes to the investigated particles caused by photon absorption are monitored.

LIR spectroscopy is one example of action spectroscopy and is performed in the Cologne laboratories regularly to investigate ions of astrophysical or chemical relevance [11–16]. It is based on the change of reaction speed or chemical equi-

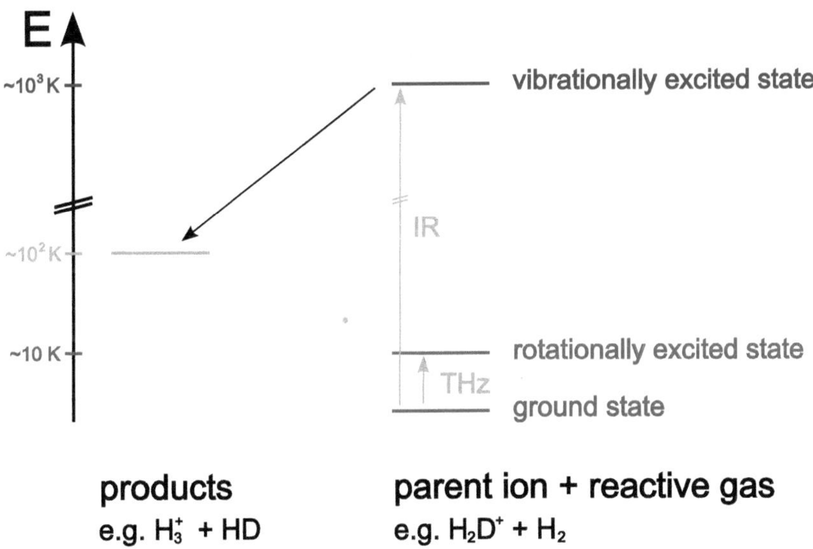

Figure 4.4: Schematic of the LIR principle. The reactants (right side) do not have enough energy to form the products (left side), when they are cooled down to the lowest possible internal state. When the ion absorbs a photon of sufficient energy, the reaction becomes possible. Endothermicities of the reactions used for LIR spectroscopy are several 100 K. By pure rotational transitions the ions gain several 10 K (THz photons), by rovibrational transitions they gain several 1000 K (IR photons). Usually pure rotational transitions do not contain enough energy to allow LIR spectroscopy.

librium in an endothermic reaction that is caused by a change of internal energy of one reaction partner due to photon absorption.

Figure 4.4 shows a schematic of a typical energy level distribution in a reaction system that can be used for LIR spectroscopy. Inserted into the trap are the reactants depicted on the right side of the figure. When cooled down to the nominal temperature of 10 K, the ions occupy only the lowest possible energy level. This is the vibrational ground state and if allowed by nuclear spin statistics (see Chapter 5), the rotational ground state. The reactants would need energies of

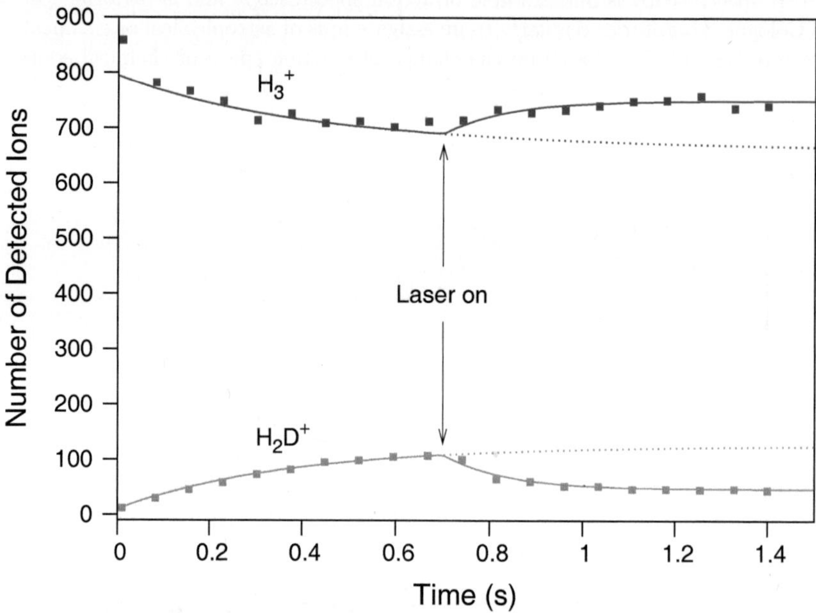

Figure 4.5: Influence of laser absorption on an endothermic reaction (from [7]). With no laser radiation present in the trap, H_3^+ reacts with HD (contained in H_2 as natural impurities) to form H_2D^+ (dotted lines). When the laser, tuned on a transition of H_2D^+, is admitted to the trap, the endothermic backward reaction of (4.6) becomes more likely and the system evolves to a different equilibrium (solid lines). The lines are numerical solutions of the rate equations for reactions (4.6) to (4.8).

The following rate coefficients and ion numbers at $t = 0$ were used for the numerical simulation: $H_{3\,0}^+ = 795$, $H_2D^+_{\,0} = 10$, $D_2H^+_{\,0} = 0$, $D_{3\,0}^+ = 0$, $k_{f1} = 1.9 \times 10^{-9}\ \frac{cm^3}{s}$, $k_{b1} = 2.9 \times 10^{-12}\ \frac{cm^3}{s}$, $k_{b1,\,Laser} = 1.4 \times 10^{-9}\ \frac{cm^3}{s}$, $k_{f2} = 0.6 \times 10^{-9}\ \frac{cm^3}{s}$, $k_{b2} = 1.0 \times 10^{-12}\ \frac{cm^3}{s}$, $k_{f3} = 0\ \frac{cm^3}{s}$, $k_{b3} = 0\ \frac{cm^3}{s}$.

usually several 100 K in order to form the products depicted on the left side of Figure 4.4. As they are cold, the probability to gain enough internal or translational energy for the reaction is extremely small. Therefore, without laser only very few reaction products can be observed.

Absorption of a photon excites the ions to higher internal levels. THz photons contain several 10 K of energy and excite rotational transitions in the vibrational ground state, while IR photons contain several 1000 K and excite rovibrational transitions (i.e. rotational and vibrational quantum numbers change). Obviously, the absorption of an IR photon immediately enables the reaction, while the absorption of a THz photon increases the probability for a reaction just slightly. Therefore, pure rotational spectroscopy is usually not possible with this method. The only exceptions are the H_2D^+ and D_2H^+ ions [17], where the endothermicity of the reaction with H_2 is comparable to the energy of the excited rotational levels in the vibrational ground state.

Figure 4.5 shows an example of laser induced reactions in the system

$$H_3^+ + HD \underset{k_{b1}}{\overset{k_{f1}}{\rightleftharpoons}} H_2D^+ + H_2 + E \tag{4.6}$$

$$H_2D^+ + HD \underset{k_{b2}}{\overset{k_{f2}}{\rightleftharpoons}} D_2H^+ + H_2 + E \tag{4.7}$$

$$D_2H^+ + HD \underset{k_{b3}}{\overset{k_{f3}}{\rightleftharpoons}} D_3^+ + H_2 + E \; . \tag{4.8}$$

For the measurement shown here, H_3^+ ions were admitted to the trap and stored together with H_2 gas, containing some HD. Reaction (4.6) evolves in the exothermic forward direction. After a storage time of 700 ms, an IR laser[6] was admitted to the trap that was tuned to an overtone transition of H_2D^+ ($0_{00} \rightarrow 1_{11}$ at 6466.532 cm^{-1} = 1546.424 nm [18]). Now the reaction system starts to evolve to another equilibrium, as the endothermic backward reaction of (4.6) becomes possible.

Figure 4.6 shows an example of the experimental setup used for LIR spectroscopy - in this case on the CH_2D^+ ion (see Chapter 7), based on the reaction system

$$CH_3^+ + HD \rightleftharpoons CH_2D^+ + H_2 + E \; . \tag{4.9}$$

From the ion source, CH_2D^+ ions were guided into the 22-pole trap. There, they were stored together with H_2 for 480 ms. During the whole trapping process, the radiation of an optical parametric oscillator (OPO, see Chapter 7.2.2) was admitted to the trap. The second quadrupole mass filter selected CH_3^+ from the

---

[6]Agilent Technologies, 81600B Tunable Laser Source Family Model #200, EC-Laser InGaAsP, wavelength range 1440 to 1640 nm, max output power <15 mW

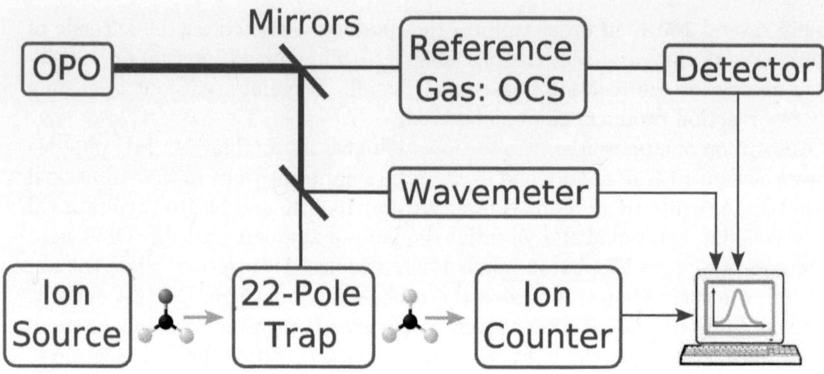

Figure 4.6: Schematic of the experimental setup for LIR spectroscopy on the example of CH_2D^+. As radiation source an OPO was used. The beam is split for simultaneous use in the 22-pole trap experiment and frequency calibration. In the ion trap, CH_2D^+ ions were stored together with H_2. Monitored were the number of reaction products CH_3^+ and the frequency of the OPO.

reaction products to be counted. Simultaneously, part of the OPO beam was used for frequency calibration measurements (see Chapter 7 for more details). Both information together result in high resolution rovibrational spectra of CH_2D^+ as shown in Figure 4.7.

The low temperatures reached with the 22-pole ion trap apparatus lead to translational temperatures of the ions of typically around 30 K. The FWHM (full-width-at-half-max) of the observed Doppler broadened transitions are therefore ≈ 0.003 cm^{-1} (the internal line widths of the radiation sources used are far below that). Thus, LIR spectroscopy is a powerful tool to obtain high resolution spectra of ions.

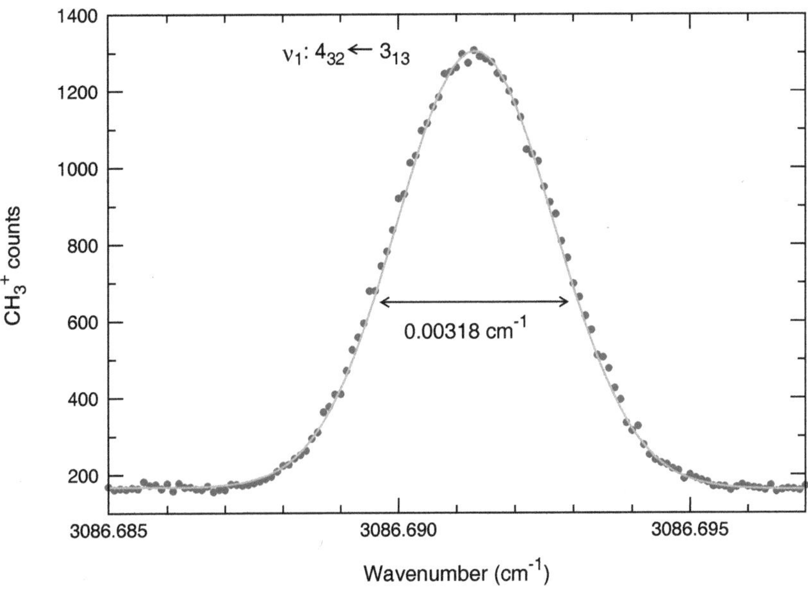

Figure 4.7: Example of a LIR line from CH_2D^+, from [13]. The absorption of
radiation is observed as an increase in the number of reaction products CH_3^+
(dots). From the Doppler width of the fitted Gaussian profile *(solid line)*, a
translational temperature of 33 K can be derived for the CH_2D^+ ions.

4.5 Daly Detectors for the New Ion Trap Experiments COLTRAP and FELion

Two new ion trap experiments have been constructed in the laboratory astrophysics group in Cologne over the last years (COLTRAP, Cologne ion trap; and
FELion, ion trap dedicated for installation at the *Free-Electron Lasers for Infrared eXperiments* (FELIX) in Nijmegen). Part of the work presented here was
the construction of two identical Daly detector systems for these experiments.

Figure 4.8 shows a schematic of the Daly principle. In the Daly detector [3],
the ions are accelerated towards an aluminium surface (knob) by a voltage of
-35 kV. They produce electrons on impact on the knob, which are accelerated
onto a scintillator (at ground potential), where they produce photons. These
photons then enter a photomultiplier tube[7] (PMT). The signal of the PMT is

---

[7]Hamamatsu PMT R647-01

then amplified[8]. A discriminator[9] allows to derive the number of ions from that signal.

A schematic of the whole detector is shown in Figure 4.9. The aluminium knob and the vacuum chamber, including the photomultiplier flange were already available. The high voltage shield and the mount for the scintillator were constructed based on earlier designs. The housing with the readout electronics was constructed within this work. Figure 4.10 shows a schematic of the electronics. Details on the construction of the housing are given in the Appendix A.3.

First operational tests of the new detectors were performed by Sven Fanghänel and are presented in his Staatsexamensarbeit [19].

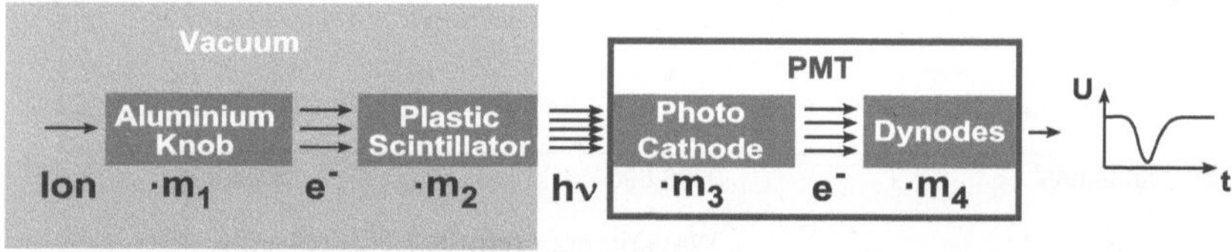

Figure 4.8: Schematic of the Daly Principle, from left to right: The ions hit an aluminium knob. Each ion produces m_1 electrons on average. These electrons are accelerated towards a plastic scintillator, where each of them produces m_2 photons. The photons leave the vacuum chamber via a window and enter the photomultiplier tube (PMT). Each photon produces m_3 electrons on the photo cathode. These electrons are then multiplied by a factor m_4 by the dynodes. The four amplification steps result in a measurable voltage signal for each ion, which is detected by an amplifier-discriminator device.

---

[8]Philips Scientific Amplifier 6950ds
[9]Philips Scientific Discriminator 6904ds

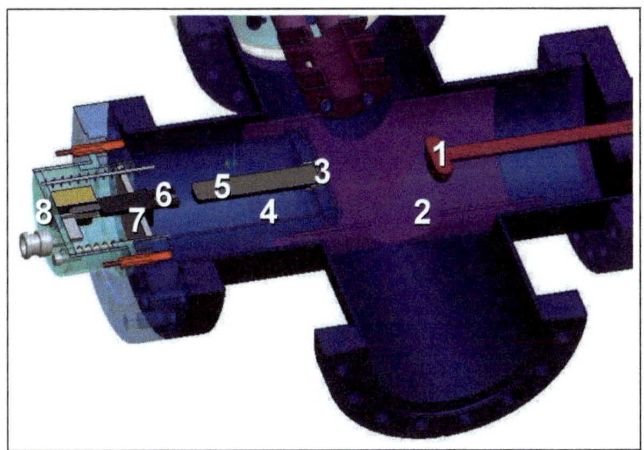

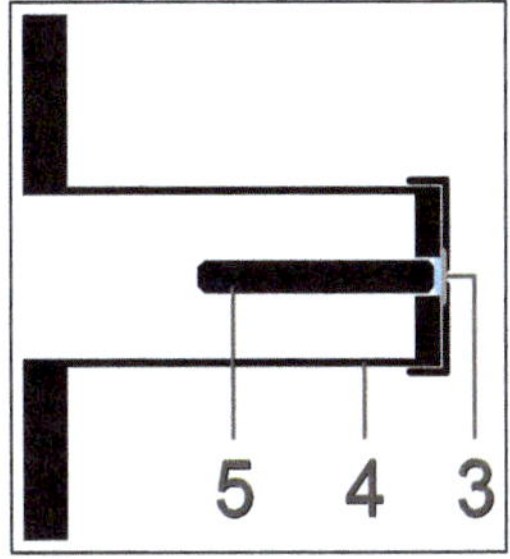

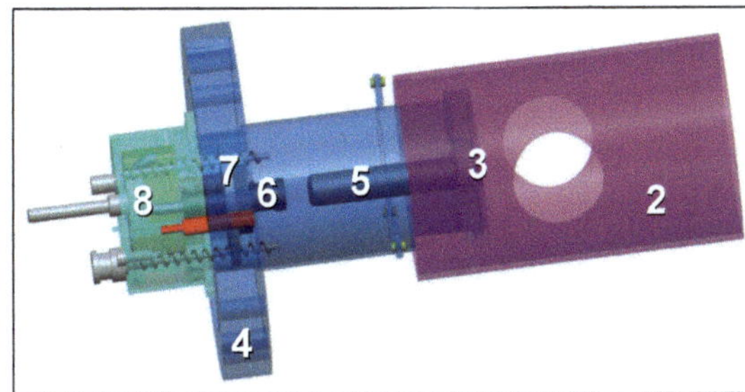

Figure 4.9: Schematic of the Daly detector. *Top:* Vacuum Chamber including all parts of the detection system (except amplifier-discriminator). *Bottom:* Photo multiplier flange. *Parts:* (from right to left)

1. Aluminium knob (-35 kV)

2. High voltage shielding tube

3. Scintillator (on the vacuum side of the window) with thin aluminium layer on top (prevents photons from the vacuum chamber from entering the PMT, lets electrons pass into the scintillator, and provides proper grounding for electrons)

4. PMT inset for vacuum chamber with window at the right end

5. Photo multiplier tube (PMT)

6. PMT socket

7. Moveable plate on springs (holds PMT socket and presses PMT to window)

8. Housing with readout and supply electronics for photomultiplier (prevents light from entering the PMT)

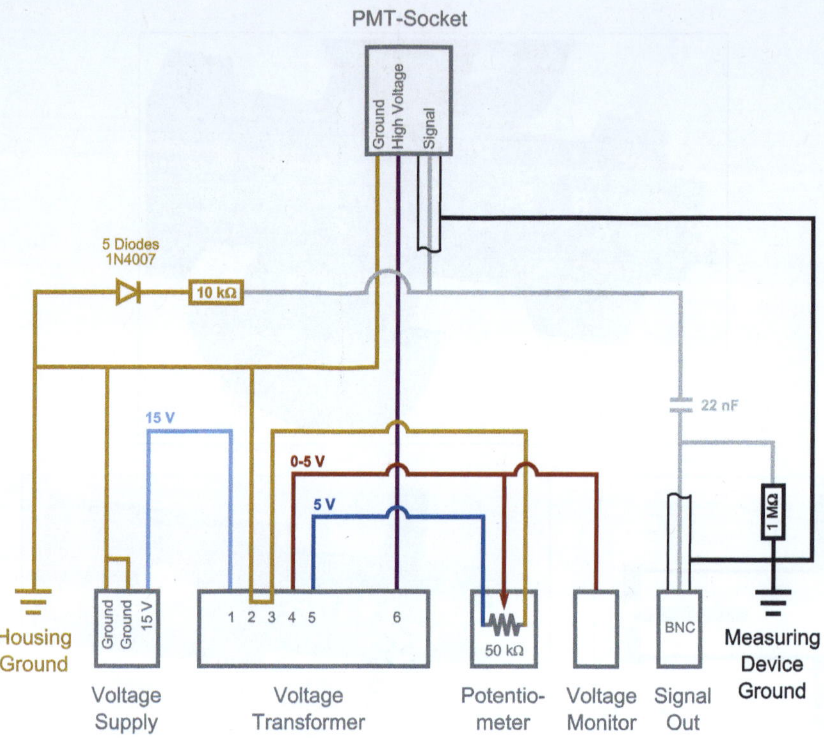

Figure 4.10: Schematic of the electronics of the new Daly detectors. The power supply yields a voltage of 15 V that is fed into the voltage transformer. The voltage converter[a] feeds 5 V into a potentiometer, from which 0 to 5 V are returned to the transformer. By tuning the potentiometer, the output voltage, that is fed to the photomultiplier can be optimized in the range of 0 to 1250 V. The output signal from the PMT is connected to the ground of the electronics housing via a set of diodes and a resistor to protect the electronics in the case of high signal voltages occurring at times, when the BNC output is not connected. The signal from the PMT is connected to the BNC output via a high-pass filter (capacitor and resistor connection to the ground of the measuring device). For the FELion apparatus, both grounds have been connected, to prevent ambient electromagnetic noise from disturbing the output signal.

---

[a]Hamamatsu, C490 Series, Compact High Voltage Power Supply Units

Chapter 5

Para Hydrogen

Molecular hydrogen can exist in two different nuclear spin configurations, named *ortho* and *para*. The differences between both, their influence on low temperature experiments and methods for the conversion from ortho to para hydrogen will be explained in Section 5.1. Methods to determine the ortho-to-para ratio in a given gas sample are described in Sections 5.2 and 5.3.

The overall wave function of a molecule contains the wave functions for the electronic and the nuclear motion. In the Born-Oppenheimer approximation these motions are considered to be independent from each other, so that the overall wave function is formed by the product of the electronic and the nuclear wave functions. Electronic transitions are not of interest here, thus, we consider only the nuclear wave function in the following. It contains the wave functions for the nuclear spin and for the vibrational and rotational motion of the nuclei. We do not consider the translation of the whole molecule here.

The hydrogen atom has a nuclear spin of $\frac{1}{2}$. Thus, when two hydrogen atoms form an H_2 molecule, there are four possible linear combinations of the nuclear spin wave functions:

$$
\begin{aligned}
|\uparrow\uparrow\rangle & \quad \text{symmetric} \\
|\downarrow\downarrow\rangle & \quad \text{symmetric} \\
|\uparrow\downarrow\rangle + |\downarrow\uparrow\rangle & \quad \text{symmetric} \\
|\uparrow\downarrow\rangle - |\downarrow\uparrow\rangle & \quad \text{antisymmetric}
\end{aligned}
\tag{5.1}
$$

Since the two hydrogen nuclei are fermions, the overall wave function must be antisymmetric under their exchange. The vibrational and electronic wave functions are symmetric in the respective ground states. Hence, the symmetry of the rotational level can only be the opposite of the symmetry of the nuclear wave function. Therefore, para hydrogen (p-H_2) can occupy only rotational levels with even values of the rotational quantum number J; ortho hydrogen (o-H_2) can only

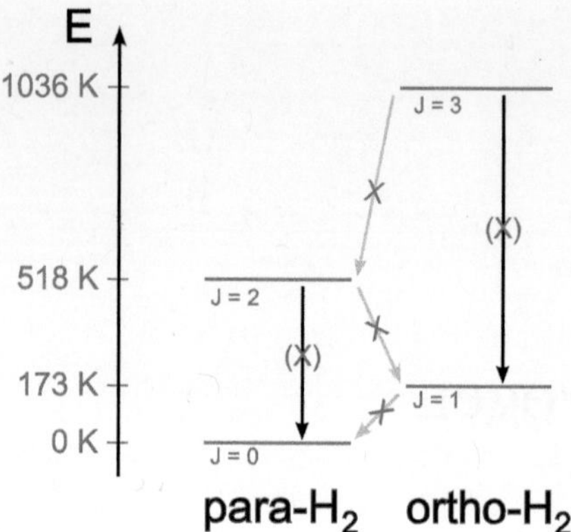

Figure 5.1: Rotational levels of molecular hydrogen. Levels with even values of the rotational quantum number J can only be occupied by p-H_2 molecules, those with J odd only by o-H_2. Radiative transitions between the two nuclear spin configurations are strictly forbidden due to conservation of the angular momentum *(grey arrows)*. The transitions marked by *black arrows* are spin allowed, but do not happen, since H_2 has no dipole moment and thus no dipole transitions.

occupy rotational levels with odd values of J (see Figure 5.1). A radiative transition between ortho and para states is strictly forbidden, as the nuclear spin must always be conserved.

The rotational levels with J odd are triply degenerate, as there are three possible symmetric linear combinations of the two nuclear spins, the J even levels are not degenerate. This leads to a ratio $\frac{\text{o-}H_2}{\text{p-}H_2} = \frac{3}{1}$ (*normal* hydrogen, n-H_2) at temperatures of ≈ 300 K or higher. Towards low temperatures, the equilibrium of the population shifts towards pure p-H_2, since the lowest possible level is a para level (see Figure 5.2). The curve in this figure is given by the thermal population of the rotational levels of H_2. Considered were the levels $J = 0$ to $J = 9$, whose energy is given by the rotational quantum number J and the rotational constant B as:

$$E_J = BJ(J+1) \tag{5.2}$$

The thermal population of each level J at a temperature T is given by

$$N_J = \frac{(2J+1)g_J}{Z} e^{-\frac{E_J}{k_B T}}, \tag{5.3}$$

where g_J is the degeneracy of level J and Z is the partition function, that cancels out for the final ratio of level occupation.

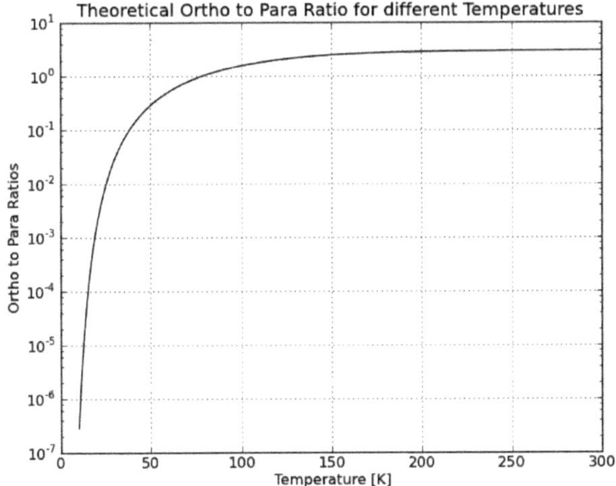

Figure 5.2: Temperature dependence of the ortho-to-para equilibrium ratio in molecular hydrogen (from [20]).

For the investigation of reactions relevant to astrochemistry, it is important to perform experiments at low temperatures, i.e. a few ten Kelvin. At these temperatures the rotational energy from the first excited rotational state often is sufficient to change the equilibrium of the observed reaction. Therefore, astrochemical experiments often need high quality para hydrogen. Ortho fractions of less than 0.1 % are desirable.

5.1 Para Hydrogen Converter

Hydrogen naturally comes as n-H_2 (75 % o-H_2, 25 % p-H_2). To perform experiments at cryogenic temperatures with p-H_2, a reliable conversion method is needed. In the Cologne laboratories, a para hydrogen generator is available for this purpose. Originally it was intended to produce a constant flow of high quality p-H_2 at very low pressures. This para hydrogen then would be used in the ion trap experiment immediately. In the following, this method is reffered to as *continuous flow production*. It will be described in Section 5.1.1.

The amounts of p-H_2 produced in continuous flow are sufficient for ion trap experiments, but not for other test methods for the ortho fraction and not for storage and later use of the gas sample. Another problem in some cases was the loss of HD compared to the natural amount in H_2 by freezing effects in the converter. As part of this work, the para generator was modified to allow the production of higher amounts of para hydrogen, which contain almost the natural abundance of deuterium. The higher amount of para hydrogen enables Raman spectroscopy as additional test method for the purity, the higher amount of deuterium makes some chemical tests of the purity in the ion trap more reliable. This new production method (*freeze out*), which is based on the operation principles presented in [21], will be described in Section 5.1.2.

In the following, it is important to keep in mind, that the temperature of the para hydrogen converter $T_{\text{converter}}$ (where the para hydrogen sample is produced) is completely independent of the temperature of the ion trap T_{trap} (where the experiments with the para hydrogen are performed).

5.1.1   Production of Para Hydrogen in Continuous Flow

For the conversion of ortho hydrogen into para hydrogen two environmental conditions are necessary: The environment has to be cooled down below 20 K and a paramagnetic catalyst material has to be present. The low temperature shifts the equilibrium of energy level occupation to almost 100 % of para hydrogen (rotational ground state). The paramagnetic catalyst enables the spin flip of the molecules from ortho to para hydrogen.

Figure 5.3 shows a schematic of the para hydrogen generator. In this case, iron oxide (Fe_2O_3[1]) powder is used as catalyst. It is filled into a copper box that is mounted onto a closed cycle helium refrigerator. For temperature control, a heating wire is attached to the cold head. Without heating, the cold head reaches a temperature of $\approx$ 12 K. At this temperature, hydrogen freezes out. The higher the heating voltage, the higher is the temperature, the cold head finally reaches. In continuous flow operation, the heating prevents hydrogen from freezing.

Around the converter box there is a cold shield. It is connected to the second cooling stage of the cold head ($\approx$ 80 K) and reduces the radiational heating of the converter box by the laboratory environment. The box is placed inside a vacuum chamber to prevent impurities like water from freezing inside the converter.

Two stainless steel tubes enter the box, one from the bottom to let H_2 in, one from the top to extract H_2. The gaskets[2] that seal the connections of the tubes to the box contain fine filters to prevent catalyst powder from leaving the box, when H_2 is flowing through the tubes and the converter box. If any catalyst would be present in warm regions of the outlet gas tube system, it would at least partially convert the produced p-H_2 back to n-H_2. Very small amounts of catalyst might fall through the filter into the bottom tube. Therefore, the p-H_2 is extracted via the top tube connection.

---

[1]Aldrich, Cat.: 31,005-0
[2]Swagelok VCR

Continuous Flow Conversion

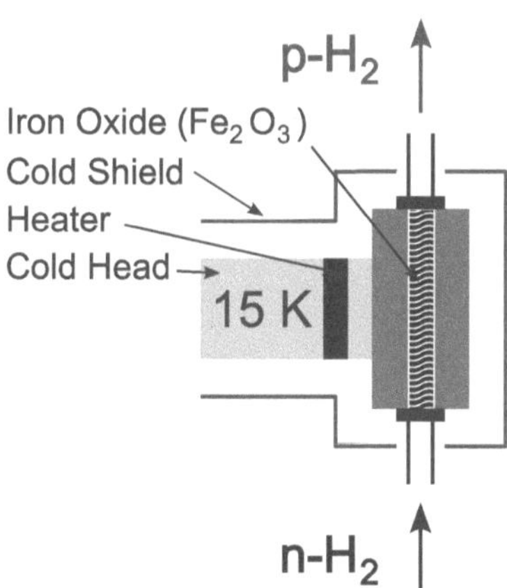

Figure 5.3: Schematic of the para hydrogen generator as used for continuous flow production of p-H$_2$. The catalyst (iron oxide powder) is kept at temperatures around 15 K. Via the bottom tube connection n-H$_2$ $\left(\frac{\text{o-H}_2}{\text{p-H}_2} = \frac{3}{1}\right)$ is flowing into the converter box. Inside the box, the ortho hydrogen molecules can attach to the catalyst and convert to para hydrogen. The para hydrogen then leaves the converter box via the top tube connection and can be used in experiments.

The para hydrogen generator used with the Cologne ion trap experiments was originally constructed to operate in continuous flow mode. In this mode, a very small amount of normal hydrogen is flowing through the cold converter box and is immediately used in the ion trap experiment (see Figure 5.3). The details on the operation of the gas tube system for this process are given in the Appendix A.4.

The method described here is sometimes difficult for experiments which investigate deuterium fractionation reactions (e.g. the H$_3^+$ isotopic system in reactions with H$_2$ isotopologues). The problems are caused by the small fraction of deuterium (HD molecules, natural HD-to-H$_2$ ratio: 3×10^{-4}), that normal hydrogen always contains. Since HD freezes out on cold surfaces preferential to H$_2$, the para hydrogen leaves the converter box with a different HD to H$_2$ ratio than the normal hydrogen had. This process reaches an equilibrium after several hours of stable operation conditions of the para hydrogen generator, but as soon as there are even slight changes to the operation conditions, such as temperature deviations of the converter box or fluctuations in the hydrogen flux, the deuterium amount of the produced para hydrogen is changing again. Additionally, the deuterium amount is always (even in the equilibrium situation) much smaller than in the normal hydrogen (HD depletion up to a factor of 60 has been observed, see Figure 5.9).

5.1.2   Production of Para Hydrogen by Freeze Out

The Cologne para hydrogen generator was modified, in order to get para hydrogen of a defined and constant quality. A gas inlet system was added, which can be used for the production of a larger amount of para hydrogen. In this method, a storage bottle is filled with hydrogen, which is then slowly frozen out into the converter box. The box is at 12 K for this purpose. After evacuating all remaining normal hydrogen from the tubes and the bottle (this usually takes roughly one hour), the converter box is slowly heated up to $\approx$ 20 K. While heating, the hydrogen evaporates from the catalyst as almost pure para hydrogen (see Figure 5.4). This method usually uses somewhat higher temperatures than the continuous flow method, leading to a small increase in the expected ortho-to-para ratio (see Table 5.1). The details of this process are given in the Appendix A.4. Figure 5.5 shows the vapour pressure curve of H_2, the pressures measured in the evaporation process tend to be slightly lower than the predicted vapour pressure for a certain temperature. This might be due to temperature variations over the converter volume during the heating process.

The freeze out method has several advantages. The most important are:

- The produced para hydrogen contains an amount of HD that is nearly as high as in the normal hydrogen.

- The HD amount of this gas stays constant over the whole experiment time.

- The amount of para hydrogen produced is sufficient to enable Raman Spectroscopy as test mechanism for the ortho-to-para ratio.

- The purity of para hydrogen stays the same during the experiment time and is not subject to any fluctuations in the production conditions, as the storage bottle is coated with Teflon, preventing a back conversion into normal hydrogen (which would be the equilibrium at room temperature).

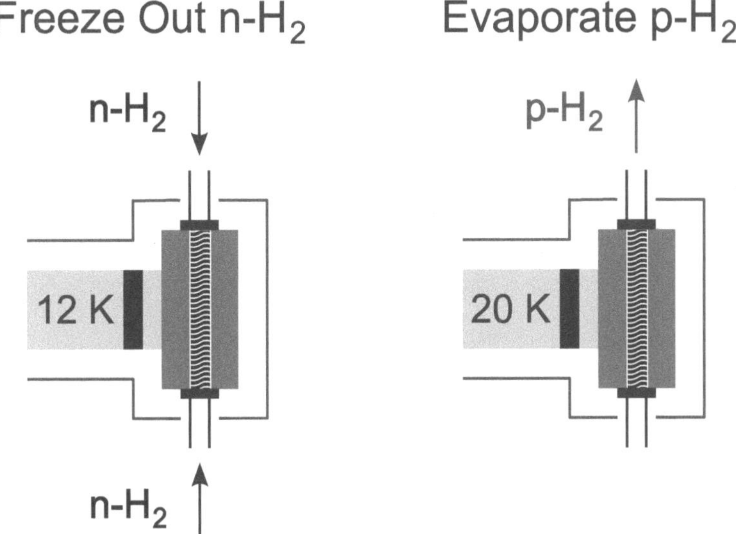

Figure 5.4: Schematic of the para hydrogen generator as used for production of p-H$_2$ by the freeze out method. First, the catalyst (iron oxide powder) is cooled down to 12 K. Via both tube connections n-H$_2$ $\left(\frac{\text{o-H}_2}{\text{p-H}_2} = \frac{3}{1}\right)$ is flowing into the converter box, where it freezes out. Inside the box, the ortho hydrogen molecules can attach to the catalyst and convert to para hydrogen. After storage times of ≈ 1 hour, the converter box is heated up to temperatures around 20 K. The para hydrogen then leaves the converter box via the top tube connection and is stored in a Teflon coated bottle for later use in experiments.

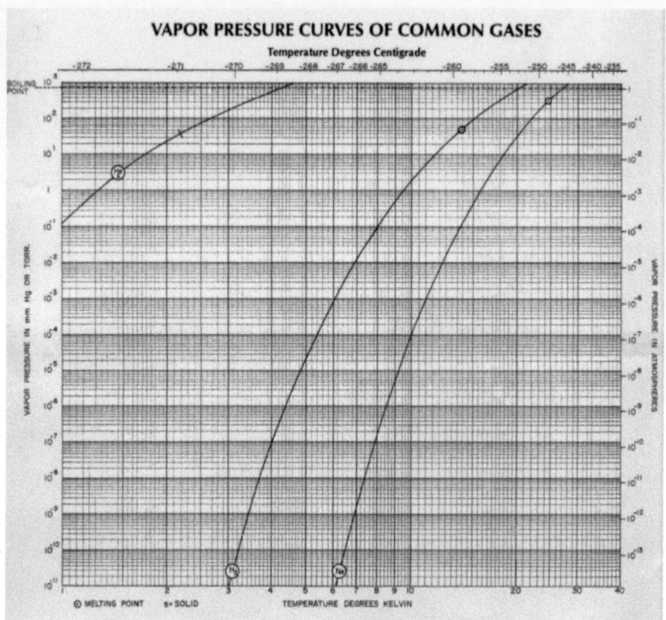

Figure 5.5: Temperature dependence of the vapour pressure of different gases, including molecular hydrogen (from [22]).

Table 5.1: Examples for the expected ortho fraction and vapour pressures at two typical operation temperatures of the para generator in the continuous flow and in the freeze out method.

Temperature (K)	$\dfrac{[\text{o-H}_2]}{[\text{o-H}_2] + [\text{p-H}_2]}$	Vapour Pressure (bar)
15	9.0×10^{-5}	0.2
20	1.6×10^{-3}	0.9

5.2 Testing the Purity of Para Hydrogen with the Reaction $\mathbf{H_3^+ + HD \rightleftharpoons H_2D^+ + H_2}$

There are several methods to determine the ortho-to-para ratio of molecular hydrogen. Direct methods like spectroscopy or nuclear magnetic resonance observe the population of rotational levels or the nuclear spin configuration. Indirect methods use the influence of the rotational energy of o-H_2 (173 K, see Figure 5.1) on chemical equilibria at low temperatures. In this work, two test reactions with ions and Raman spectroscopy were used.

In the ion trap apparatus, the investigation of the ortho-to-para ratio of H_2 is only possible indirectly. There are several ions whose reactions with H_2 are only slightly endothermic. In such a case the rotational energy of o-H_2 is sufficient to enable the reaction, while p-H_2 does usually not contain enough energy for a reaction at cryogenic temperatures. Therefore, the observation of such chemical equilibria allows to determine the ortho-to-para ratio of the reacting hydrogen.

Suitable reactions are for example:

- $H_3^+ + H_2 \rightleftharpoons H_3^+ + H_2$: Similar to H_2, H_3^+ can exist in either ortho (total nuclear spin $\frac{3}{2}$) or para (total nuclear spin $\frac{1}{2}$) states. The nuclear spin configuration determines which rovibrational levels the ion can occupy and thus can be probed spectroscopically [23, 24]. The ortho-to-para ratio of the reactant H_2 molecules determines the ortho-to-para ratio of H_3^+ in the thermal reaction equilibrium. Therefore, this reaction can serve as a probe for the purity of a p-H_2 sample. However, it has not been applied for this purpose in the present work and will not be discussed in further detail here.

- $H_2D^+ + H_2 \rightleftharpoons H_3^+ + HD$: This reaction is endothermic in the forward direction. Thus its forward rate is enhanced by the rotational energy of o-H_2. The dependence of the reaction equilibrium on the ortho-to-para ratio of H_2 is a very sensitive probe for this ratio. Measurements on this reaction system will be presented in the following.

- $N^+ + H_2 \rightleftharpoons NH^+ + H$: This reaction is also endothermic in the forward direction and can be used as probe for the ortho-to-para ratio of H_2 as well. So far, it is not known what influence the fine-structure energy of the N^+ ions has on the reaction equilibrium. Investigations towards a more detailed understanding of this reaction system are presented in Chapter 6.

5.2.1 H_3^+ Isotopologues

The isotopologues of H_3^+ are the main drivers of deuterium fractionation in cold molecular clouds via the reactions

$$H_3^+ \quad + HD \; \rightleftharpoons \; H_2D^+ + H_2 + 232 \text{ K} \tag{5.4}$$

$$H_2D^+ + HD \; \rightleftharpoons \; D_2H^+ + H_2 + 187 \text{ K} \tag{5.5}$$

$$D_2H^+ + HD \; \rightleftharpoons \; D_3^+ \quad + H_2 + 234 \text{ K} . \tag{5.6}$$

Therefore, their reactions and spectroscopy have been investigated thoroughly[3]. As shown in Figure 4.5, the equilibrium of reaction (5.4) depends on the kinetic energy of the reaction partners on the right side. When one of them gains enough translational, rotational or vibrational energy to overcome the endothermicity, the back reaction becomes possible. This energy might be gained thermally, in this case the probability is given by the Boltzmann distribution for the temperature of the system. The required energy might also be gained from laser irradiation or it might be determined by external parameters like nuclear spin statistics.
The forward reaction of (5.4) needs HD, so the equilibrium of the reaction depends on the temperature (translational energy), the ortho-to-para ratio of H_2 (rotational energy) and the HD-to-H_2 ratio.

5.2.2   Theory

Any reaction system that can be described as

$$A + B \; \underset{k_b}{\overset{k_f}{\rightleftharpoons}} \; C + D ,$$

is in its equilibrium, when the forward reaction rate and the back reaction rate are equal:

$$k_f[A][B] \;=\; k_b[C][D]$$

$k_{f/b}$ denotes the rate coefficient for forward/back reaction, [X] is the number density of species X. For reaction (5.4) this means:

$$k_f[H_3^+][HD] \;=\; k_b[H_2D^+][H_2]$$

---

[3]Isotopic System: [9, 25, 26]
   Reactions: [27, 28]
   Spectroscopy: [17, 18, 29–42]
   Interstellar: [43–47]

The simplest approach to include different nuclear spin configurations explicitly, is to treat them as different species

$$H_3^+ + HD \quad \underset{k_{b,pp}}{\overset{k_{f,pp}}{\rightleftharpoons}} \quad \text{p-}H_2D^+ + \text{p-}H_2 + 232\ K \tag{5.7}$$

$$H_3^+ + HD \quad \underset{k_{b,op}}{\overset{k_{f,op}}{\rightleftharpoons}} \quad \text{o-}H_2D^+ + \text{p-}H_2 + 146\ K \tag{5.8}$$

$$H_3^+ + HD \quad \underset{k_{b,po}}{\overset{k_{f,po}}{\rightleftharpoons}} \quad \text{p-}H_2D^+ + \text{o-}H_2 + 59\ K \tag{5.9}$$

$$H_3^+ + HD + 25\ K \quad \underset{k_{b,oo}}{\overset{k_{f,oo}}{\rightleftharpoons}} \quad \text{o-}H_2D^+ + \text{o-}H_2 \tag{5.10}$$

Gerlich et al [25] follow this approach. They include the nuclear spins of H_2 and H_2D^+ (o-H_2D^+ has a rotational energy of 86 K in its lowest state $J = 1$). The influence of the nuclear spin of the H_3^+ ions, of rotational levels with $J > 1$ for any reaction partner and interactions between H_2D^+ and H_2 (see [9]) are neglected, although the nuclear spin configuration of the reacting H_3^+ ions will influence the nuclear spin configurations of the products (see [23, 24]). However, in this model we can assume, that the exothermic reactions happen all at the same rate k'. The rate for the endothermic direction is then k' multiplied with the probability to gain the required energy thermally :

$$k_{f,pp} = k_{f,op} = k_{f,po} = k_{b,oo} = k' \tag{5.11}$$

$$k_{b,pp} = k' \cdot e^{-\frac{232\ K}{T}} \tag{5.12}$$

$$k_{b,op} = k' \cdot e^{-\frac{146\ K}{T}} \tag{5.13}$$

$$k_{b,po} = k' \cdot e^{-\frac{59\ K}{T}} \tag{5.14}$$

$$k_{f,oo} = k' \cdot e^{-\frac{25\ K}{T}} \tag{5.15}$$

T is the translational temperature of the particles. For simplicity, we treat each reaction as independent from the others, although they all start with the same H_3^+ ions and HD molecules. Hence normalization factors would be needed to be generally correct. However, in the simple model the equilibria of reactions (5.7) to (5.10) are given by:

$$k' \quad [H_3^+][HD] = k' \cdot e^{-\frac{232\ K}{T}} [\text{p-}H_2D^+][\text{p-}H_2] \tag{5.16}$$

$$k' \quad [H_3^+][HD] = k' \cdot e^{-\frac{146\ K}{T}} [\text{o-}H_2D^+][\text{p-}H_2] \tag{5.17}$$

$$k' \quad [H_3^+][HD] = k' \cdot e^{-\frac{59\ K}{T}} [\text{p-}H_2D^+][\text{o-}H_2] \tag{5.18}$$

$$k' \cdot e^{-\frac{25\ K}{T}} [H_3^+][HD] = k' \quad [\text{o-}H_2D^+][\text{o-}H_2] \tag{5.19}$$

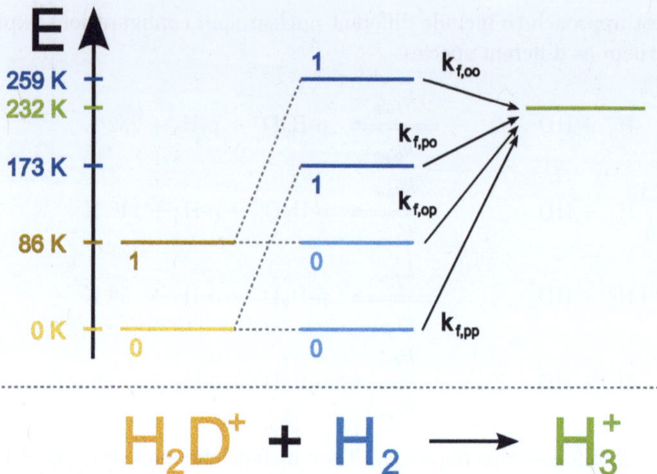

Figure 5.6: Scheme of state-specific rate coefficients for the reaction system $H_3^+ + HD \rightleftharpoons H_2D^+ + H_2$.
The two lowest rotational levels ($J = 0, 1$) of H_2D^+ *(yellow)* can react with the two lowest rotational levels ($J = 0, 1$) of H_2 *(blue)* to form H_3^+ *(green)*. Each combination of levels has another rate coefficient k_f for the reaction. The backward rate coefficients are not shown for simplicity.

The equilibrium of the total reaction system is the sum of Equations (5.16) to (5.19).

$$[H_3^+][HD]\left(1 + 1 + 1 + e^{-\frac{25\,K}{T}}\right) = e^{-\frac{232\,K}{T}}[\text{p-}H_2D^+][\text{p-}H_2] + e^{-\frac{146\,K}{T}}[\text{o-}H_2D^+][\text{p-}H_2]$$
$$+ \; e^{-\frac{59\,K}{T}}[\text{p-}H_2D^+][\text{o-}H_2] + [\text{o-}H_2D^+][\text{o-}H_2]$$

$$(5.20)$$

From this we can calculate the $\frac{[H_2D^+]}{[H_3^+]}$ ratio by introducing the ortho fractions of H_2 and H_2D^+, $f = \frac{[\text{o-}H_2]}{[H_2]}$ and $g = \frac{[\text{o-}H_2D^+]}{[H_2D^+]}$:

$$\frac{[H_2D^+]}{[H_3^+]} = \frac{[HD]}{[H_2]} \frac{1 + 1 + 1 + e^{-\frac{25\,K}{T}}}{e^{-\frac{232\,K}{T}}(1-g)(1-f) + e^{-\frac{146\,K}{T}}g(1-f) + e^{-\frac{59\,K}{T}}(1-g)f + gf}$$

$$(5.21)$$

In order to determine f from the observed $\frac{[H_2D^+]}{[H_3^+]}$ ratio, we need to know the value of g. One possible approach is to derive it from Equations (5.16) to (5.19).

$$\frac{[\text{o-}H_2D^+]}{[\text{p-}H_2D^+]} = \frac{1 + e^{-\frac{25\ K}{T}}}{2} \cdot \frac{e^{-\frac{232\ K}{T}}(1-f) + e^{-\frac{59\ K}{T}}f}{e^{-\frac{146\ K}{T}}(1-f) + f} \tag{5.22}$$

For a temperature of 10 K and values of $f > 10^{-5}$, which are reasonable assumptions for our experiments, this becomes $\approx 3 \times 10^{-3}$. Equation (5.22) is based on the assumption that the ortho-to-para ratio of H_2D^+ is given only by the equilibria of reactions (5.7) to (5.10). As most of those reactions are endothermic, a collisional complex built by H_2D^+ and H_2 will much more likely separate as H_2D^+ and H_2 again instead of forming H_3^+ and HD. In such collisions, the nuclear spins of the partners can be changed as long as the total nuclear spin of the partners is conserved. As explained in [25] the following nuclear spin conversion reactions are allowed:

$$\text{o-}H_2D^+ + \text{p-}H_2 \leftrightharpoons \text{p-}H_2D^+ + \text{o-}H_2 \tag{5.23}$$
$$\text{o-}H_2D^+ + \text{o-}H_2 \leftrightharpoons \text{p-}H_2D^+ + \text{o-}H_2 \tag{5.24}$$
$$\text{o-}H_2D^+ + \text{o-}H_2 \leftrightharpoons \text{p-}H_2D^+ + \text{p-}H_2 \tag{5.25}$$

Considering these reactions and assuming a nominal temperature of 10 K, Gerlich et. al. [25] derived the relation:

$$\frac{[\text{o-}H_2D^+]}{[\text{p-}H_2D^+]} = \frac{f}{2.2 \times 10^{-4} + f} \tag{5.26}$$
$$\Rightarrow \quad g = \frac{f}{2.2 \times 10^{-4} + 2f} \tag{5.27}$$

Their approach also includes the fact, that in the ion trap experiments H_2 is flowing through the trap continuously and thus f can be considered a constant. For reasonable values of f ($> 10^{-5}$), (5.26) varies between 0.05 and 1. Thus, the influence of collisions with H_2 cannot be neglected for the ortho fraction of H_2D^+.

With (5.27), Equation (5.21) becomes:

$$
\begin{aligned}
\frac{[\mathrm{H_2D^+}]}{[\mathrm{H_3^+}]} \;=\;& \frac{[\mathrm{HD}]}{[\mathrm{H_2}]} \left(3 + e^{-\frac{25\,\mathrm{K}}{T}}\right)\left(e^{-\frac{232\,\mathrm{K}}{T}}\left(2.2 \times 10^{-4} + \left(1 - 2.2 \times 10^{-4}\right)f + f^2\right)\right. \\[2ex]
&\left. + \; e^{-\frac{146\,\mathrm{K}}{T}}\left(f - f^2\right) + e^{-\frac{59\,\mathrm{K}}{T}}\left(2.2 \times 10^{-4}f + f^2\right) + f^2\right)^{-1}\left(2.2 \times 10^{-4} + 2f\right) \\[2ex]
\;\approx\;& \frac{[\mathrm{HD}]}{[\mathrm{H_2}]}\left(3 + e^{-\frac{25\,\mathrm{K}}{T}}\right)\left(2.2 \times 10^{-4} + 2f\right) \\[2ex]
&\cdot \; \left(e^{-\frac{232\,\mathrm{K}}{T}}\left(2.2 \times 10^{-4} + f\right) + e^{-\frac{146\,\mathrm{K}}{T}}f + e^{-\frac{59\,\mathrm{K}}{T}}\left(2.2 \times 10^{-4}f + f^2\right) + f^2\right)^{-1}
\end{aligned}
$$

$$\tag{5.28}$$

In the last step, we assumed that $\left(1 - 2.2 \times 10^{-4}\right) \approx 1$ and $f \pm f^2 \approx f$, since f is supposed to be small. With (5.28) we now have an analytic expression for the $\frac{[\mathrm{H_2D^+}]}{[\mathrm{H_3^+}]}$ ratio that will allow us to derive the ortho fraction f of the reactant hydrogen from measurements where the temperature T and the $\frac{[\mathrm{HD}]}{[\mathrm{H_2}]}$ ratio are known.

5.2.3   Measurements in Continuous Flow Mode

A series of measurements was performed in order to determine the o-$\mathrm{H_2}$ fraction remaining in the hydrogen, when it is converted to p-$\mathrm{H_2}$ in the para hydrogen generator in continuous flow operation. As described in Section 5.1.1, this means, that a constant flux of n-$\mathrm{H_2}$ is flowing into the cold converter box, where it is converted to p-$\mathrm{H_2}$. The para hydrogen is then lead to the ion trap, where it is used for experiments immediately. The temperature of the converter box was 18.8 K, the nominal temperature of the 22-pole trap was 12.8 K.

Figure 5.7 shows the first and the last measurement of this series. The resulting ratios $\frac{[\mathrm{H_2D^+}]}{[\mathrm{H_3^+}]}$ are shown in Figure 5.8. The change of this ratio with the operation time of the para generator is to be expected as explained in Section 5.1.1.

Equation (5.28) gives an expected $\frac{[\mathrm{H_2D^+}]}{[\mathrm{H_3^+}]}$ ratio of 2.2 at the measurement conditions, while the results in Figure 5.8 are around 9×10^{-3}. So the question was, how to explain a discrepancy of a factor of ≈ 250.

- As mentioned before, the model leading to Equation (5.28) is simplified and surely contains some errors itself, but not that much.

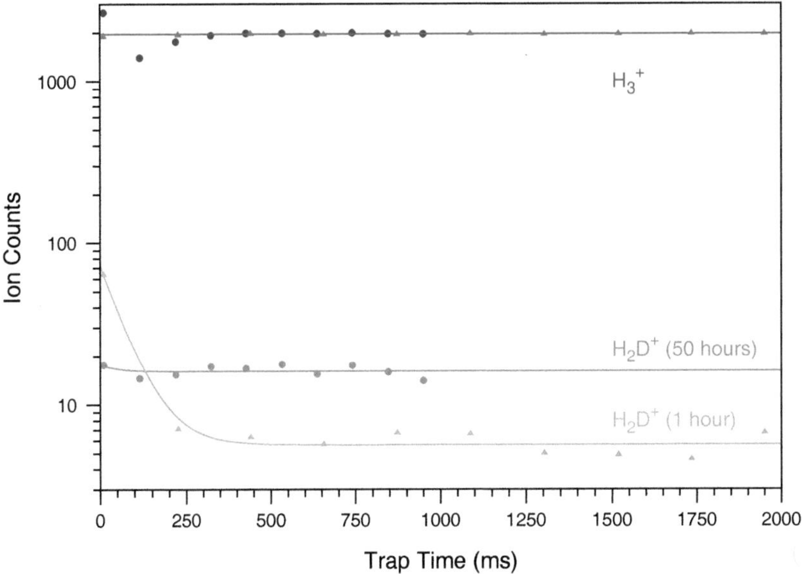

Figure 5.7: Time evolution of the reaction system $H_3^+ + HD \rightleftharpoons H_2D^+ + H_2$. The para hydrogen generator was operated in continuous flow mode. Two measurements are shown, after 1 hour *(triangles)* and after 50 hours *(circles)* operation time of the para generator.

The equilibrium amount of H_2D^+ for later measurements is higher than for the early ones. This is most likely due to the reduced HD amount in the p-H_2 after short operation times of the para generator. The p-H_2 was produced at a converter temperature of 18.8 K for these measurements, leading to an expected ortho fraction of 9.23×10^{-4} in the reactant gas.

- Earlier tests [7] showed, that the temperature of the neutral gas is usually a few Kelvin higher than the nominal temperature of the ion trap. Including this effect improves the results by less than a factor of two (see Table 5.2). Thus, it is neglected in the following.

- Also, the $\frac{[o\text{-}H_2]}{[H_2]}$ fraction might deviate from the thermal equilibrium at 18.8 K, resulting in about one order of magnitude.

- A possible explanation for the remaining deviations likely is a lack of HD, which will be discussed in the following.

The $\frac{[HD]}{[H_2]}$ equilibrium ratio in the p-H_2 sample can be significantly lower than the natural ratio. In this case, the $\frac{[\mathrm{H_2D^+}]}{[\mathrm{H_3^+}]}$ ratio in our measurements decreases by the same factor, as can be seen from Equation (5.21). This effect is visible also in Figure 5.8, where the $\frac{[\mathrm{H_2D^+}]}{[\mathrm{H_3^+}]}$ ratio increased by a factor of ≈ 3.6 from $t = 0$ h to $t = 20$ h, due to changes in the HD amount by freezing effects.

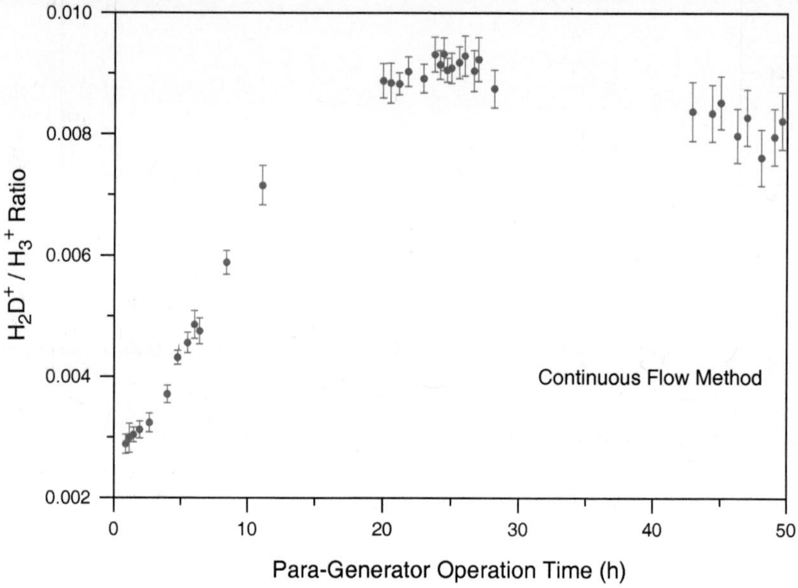

Figure 5.8: The ratio $\frac{[\mathrm{H_2D^+}]}{[\mathrm{H_3^+}]}$ changes with the time, the para hydrogen generator is operated in continuous flow mode. On cold and clean surfaces, the sticking probability is much higher for HD than for H$_2$. This leads to a reduced amount of HD in the produced p-H$_2$. As the surface coverage with HD increases, the amount of HD in the evaporated gas increases, leading to a higher equilibrium amount of H$_2$D$^+$. This process reaches an equilibrium after operation times of ≈ 20 hours. At that time, the $\frac{[\mathrm{H_2D^+}]}{[\mathrm{H_3^+}]}$ ratio and thus the $\frac{[\mathrm{HD}]}{[\mathrm{H_2}]}$ ratio increased by a factor of ≈ 3.6 w.r.t. to the value at $t = 0$. Any shifts in the production conditions of p-H$_2$ also change the $\frac{[\mathrm{H_2D^+}]}{[\mathrm{H_3^+}]}$ ratio, as can be seen for the long operation times.

A quick method to get an idea of the compounds of a gas, is to ionize it in the ion source and take a mass spectrum of the ionization products. This was done with n-H$_2$ and with the p-H$_2$ coming from the para generator. Together with the natural amounts of HD, H$_2$ forms all possible isotopologues of H$_3^+$, when ionized and stored in the ion source. A lack in HD results in a reduced number of deuterated isotopologues after ionization of the gas. The results are shown in Figure 5.9. It turned out, that the $\frac{[H_2D^+]}{[H_3^+]}$ ratio in the ions coming from the source is three orders of magnitude lower for p-H_2 than for n-H_2.

From these measurements we can derive estimates of the $\frac{[HD]}{[H_2]}$ ratio. For the measurement shown in Figure 5.9, the $\frac{[H_2D^+]}{[H_3^+]}$ ratio was decreased by a factor of $\frac{1}{447}$, thus we assume the $\frac{[HD]}{[H_2]}$ ratio to be decreased by the same factor, leading to a value of 6.71×10^{-7} for a para generator operation time of two hours. For operation times of the para generator longer than 20 h, we expect this value to increase by a factor of three, leading to an $\frac{[HD]}{[H_2]}$ ratio of 2.01×10^{-6}. We insert this and the nominal trap temperature of 12.8 K into Equation (5.28) and then vary the assumed value of the $\frac{[o\text{-}H_2]}{[H_2]}$ fraction until the predicted $\frac{[H_2D^+]}{[H_3^+]}$ ratio becomes 9×10^{-3}. The best fitting ortho fraction for these assumptions is 1.48×10^{-3} which is quite close to the thermal equilibrium of 9.23×10^{-4} at 18.8 K (compare Figure 5.11). If we assume the thermal equilibrium value for the ortho fraction, we can use Equation (5.28) to derive the HD amount of the reactant gas to be 1.21×10^{-6}. Those two results can serve as upper and lower limits for the ortho fraction, but a reasonable estimation of errors for them is not possible. However, all results are summarized in Table 5.2.

5.2.4 Measurements in Freeze Out Mode

The measurements on the purity of para hydrogen presented so far revealed some difficulties. Especially the fact that two parameters that influence the $\frac{[H_2D^+]}{[H_3^+]}$ ratio are unknown (the $\frac{[o\text{-}H_2]}{[H_2]}$ fraction and the $\frac{[HD]}{[H_2]}$ ratio), calls for at least one more test method. The idea is, to have a set of three test methods that are calibrated against each other to determine those two parameters. Those test methods are the already explained reaction system $H_2D^+ + H_2$, the system $N^+ + H_2$ and Raman spectroscopy. The reaction system $N^+ + H_2$ is very interesting on its own. Measurements towards the full understanding of this system will be presented in Chapter 6. Raman spectroscopy will be explained in Section 5.3. As Raman spectroscopy needs much higher amounts of gas than test reactions in the 22-pole ion trap, the para generator had to be modified. In order to produce some hundred millibar of p-H_2, the normal hydrogen has to be frozen out into the converter and then evaporated at temperatures around 20 K into a storage bottle.

The higher evaporation temperatures of this method and the fact, that there is a fixed amount of both gases (H_2 and HD) in the converter box when the evaporation is started might lead to amounts of HD in the produced p-H_2 that are much closer to the natural abundance. In order to test this method, the para generator was modified in a temporary way. A stainless steel gas reservoir was used as storage bottle for the p-H_2 and the first p-H_2 was produced by

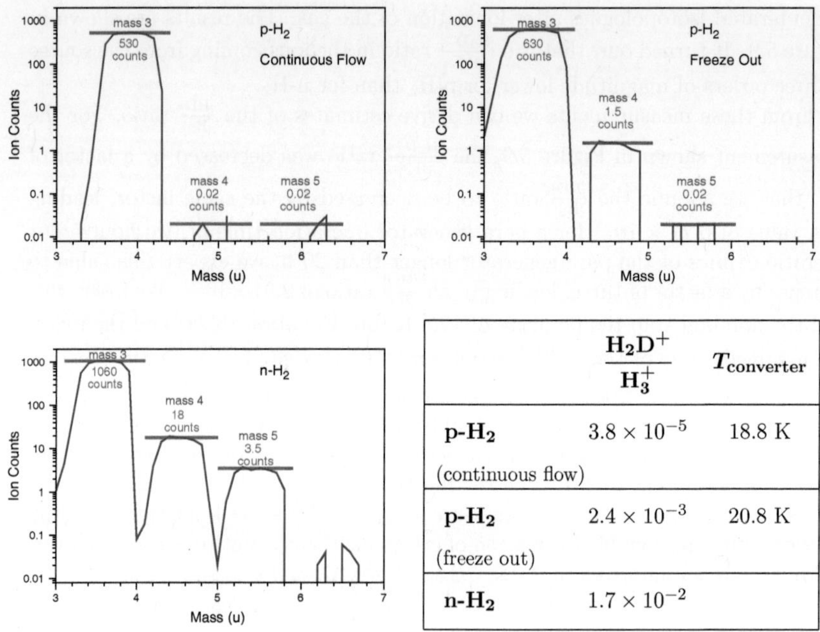

	$\dfrac{\mathbf{H_2D^+}}{\mathbf{H_3^+}}$	$T_{\mathbf{converter}}$
p-H₂ (continuous flow)	3.8×10^{-5}	18.8 K
p-H₂ (freeze out)	2.4×10^{-3}	20.8 K
n-H₂	1.7×10^{-2}	

Figure 5.9: Mass spectra of ionization products from n-H₂ and p-H₂. The ionization of p-H₂ shows reduced amounts of deuterated H_3^+ isotopologues compared to the ionization of n-H₂. The production of p-H₂ in continuous flow reduces the available deuterium amount even more, than the freeze out production. For these measurements, the ions were not stored in the 22-pole trap and the shortest possible cycle time of 50 ms was chosen. For the measurements in continuous flow mode, the para generator has been operated since two hours. Thus, the equilibrium ratio of $\frac{H_2D^+}{H_3^+}$ can be expected to be reduced by a factor of ≈ 3 w.r.t. to the long time equilibrium for this production method (compare Fig. 5.8).

the new method with the maximum evaporation temperature of 20.8 K (the heating voltage was increased slowly, until the desired pressure of 0.6 bar was reached). Comparison with Figure 5.5 reveals, that this point lies below the vapour pressure curve of H₂, which can be taken as a hint that the average temperature of the converter is somewhat below the temperature read from the silicon diode. Figure 5.9 shows the ionization products from n-H₂ in comparison with those from p-H₂ by both production methods. The table in this figure gives the $\frac{[H_2D^+]}{[H_3^+]}$ ratios produced in the ion source. The HD amount is much closer to the natural ratio for the freeze out method, than it was for the continuous flow production.

The measurements on the $\frac{[H_2D^+]}{[H_3^+]}$ ratio were repeated with the p-H_2 sample produced by the freeze out method (Figure 5.10). Two effects can be seen in the time evolution of this equilibrium ratio. One is the decrease with the storage time of p-H_2 in the stainless steel bottle (green function in Figure 5.10). This is due to an increase of the o-H_2 fraction with the storage time, as conversion processes can happen on the surfaces of the bottle. The other effect is an increase

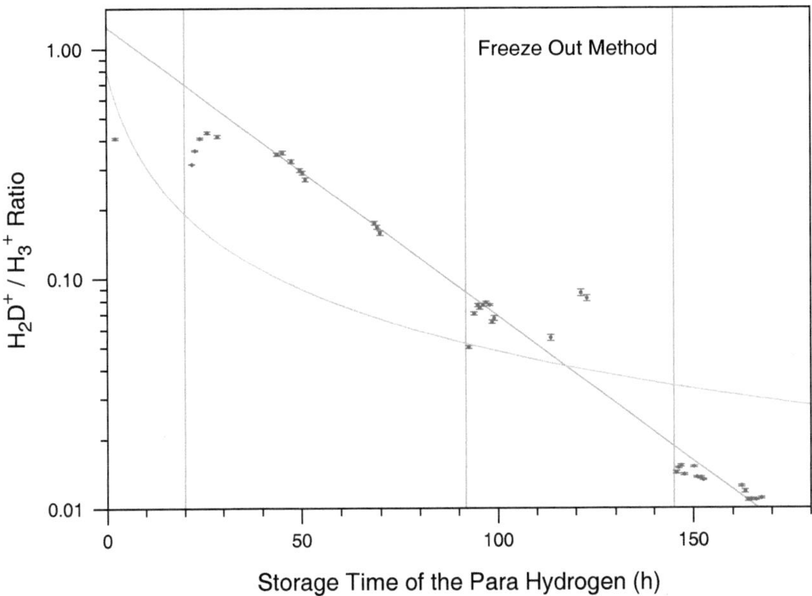

Figure 5.10: The ratio $\frac{[H_2D^+]}{[H_3^+]}$ changes with the time, the para hydrogen is stored in a stainless steel bottle. On the surfaces of the bottle, some of the p-H_2 converts to o-H_2, which has a higher probability to react with H_2D^+ to H_3^+, so that the ratio $\frac{[H_2D^+]}{[H_3^+]}$ decreases. The *vertical lines* mark the times, when the ion trap apparatus was heated up over night and measurements started with clean surfaces in the next morning, leading to preferential freeze out of HD and thus to a reduced $\frac{[H_2D^+]}{[H_3^+]}$ ratio. The *straight line* shows function (5.30) with a time constant of 34.5 h, the *bend curve* shows the result of inserting Equation (5.29) into Equation (5.28) with a time constant of 2000 h, assuming the natural abundance of HD and the thermal equilibrium ratio at 20.8 K of 2.4×10^{-3} for the ortho fraction.

in the $\frac{[H_2D^+]}{[H_3^+]}$ ratio with time towards the values on the green curve, any time the trap was heated up (*light blue lines* in Figure 5.10). This is due to the already mentioned preferential freezing of HD. When the trap was heated up over night, the next low temperature measurements started with clean surfaces. Hence, the $\frac{HD}{H_2}$ ratio was very low in the beginning and evolved towards its equilibrium over timescales of five to ten hours. The equilibrium is reached on shorter timescales than in the para generator, because in the trap there is much less surface to cover.

To describe the dependence of the $\frac{[H_2D^+]}{[H_3^+]}$ ratio on the storage time of p-H$_2$ in the gas reservoir, we need to insert the time dependence of o-H$_2$ into Equation (5.28):

$$f(t) = \tfrac{3}{4} - \left(\tfrac{3}{4} - f_0\right) e^{-\frac{t}{\tau}} \qquad (5.29)$$

where $\tfrac{3}{4}$ is the equilibrium ratio of $f = \frac{[\text{o-H}_2]}{[\text{H}_2]}$ at room temperature, f_0 the ratio at $t = 0$, and τ is the time constant for back conversion. However, it is not possible to describe the measurement results shown in Figure 5.10 by the resulting function. Instead, the decrease of the $\frac{[H_2D^+]}{[H_3^+]}$ ratio with the p-H$_2$ storage time can be described by the much simpler function

$$\frac{[H_2D^+]}{[H_3^+]}(t) = (1.25 \pm 0.10) \cdot \exp\left(-\frac{t}{(34.5 \,\pm\, 2.0)\ \text{h}}\right) . \qquad (5.30)$$

This can be interpreted as an $\frac{[H_2D^+]}{[H_3^+]}$ ratio at $t = 0$ of 1.25 ± 0.10 and a time constant for the conversion of p-H$_2$ to o-H$_2$ of (34.5 ± 2.0) h. With the best guess for the $\frac{HD}{H_2}$ ratio of 4.2×10^{-5} (see Figure 5.9), this leads to an $\frac{[\text{o-H}_2]}{[\text{H}_2]}$ fraction of 2.78×10^{-4}. This value is even below the thermal equilibrium of 2.23×10^{-3} at the maximum evaporation temperature of 20.8 K. This is to be expected from the production process. Since the hydrogen evaporates from the converter box during the whole heating process, the $\frac{[\text{o-H}_2]}{[\text{H}_2]}$ fraction should be below the thermal equilibrium of the maximal temperature. The ortho fraction of 2.78×10^{-4} corresponds to a temperature of 16.6 K, where the vapour pressure of H$_2$ is less than half of that at 20.8 K. Thus, this ortho fraction is probably too low. If we consider the natural HD amount instead of the reduced one, we can derive an ortho fraction of 1.58×10^{-3} which corresponds to a temperature of 19.8 K (compare Figure 5.11). Since the measurements shown in Figure 5.9 clearly indicate a reduced HD amount, this ortho fraction is surely too high. As before, the two values yield the upper and lower limit for the ortho fraction.

The measurements presented here were not yet repeated with the final setup of the para hydrogen generator. Compared to the Raman results for the glass cell

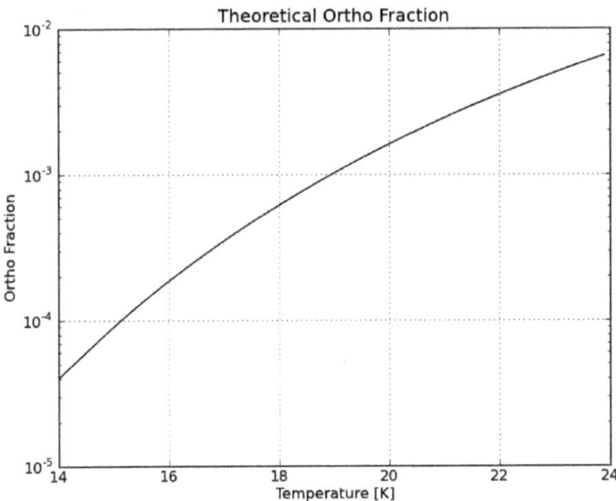

Figure 5.11: Temperature dependence of the ortho-to-para ratio in molecular hydrogen in the range of 14 to 24 K (based on [20]).

(see Section 5.3), the rate for the back conversion of p-H_2 is rather large. This might be explained by rust on the inner surfaces of the gas reservoir. However, the simplified model used here (inserting Equation (5.29) in Equation (5.28)) clearly reached its limits in the description of the influence of back conversion on the $\frac{[H_2D^+]}{[H_3^+]}$ ratio. Thus, we cannot be sure that the fitted time constant of 34.5 h does not include other effects besides the back conversion.

For future measurements, the ion trap should be operated as long as possible at cryogenic temperatures to reduce effects of the HD freezing. Preferably, p-H_2 should be flowing through the ion trap for several hours, even before the measurements are started with a fresh p-H_2 sample. That way the results for short storage times get more reliable. The Teflon coating of the new storage bottles will most likely lead to longer back conversion times.

The description of the dependence of the $\frac{[H_2D^+]}{[H_3^+]}$ ratio on the storage time of the hydrogen sample cannot be improved so easily. The complete model for the equilibrium $\frac{[H_2D^+]}{[H_3^+]}$ ratio requires the explicit treatment of all possible internal states of all involved reaction partners (see [26]). However, the results presented here are strong hints, that the ortho fractions in the p-H_2 samples are close to the thermal equilibrium at the converter temperature and that the HD amounts of the samples are lower than the natural abundance, especially for the continuous flow method.

All results on the purity of para hydrogen will be summarized in Section 6.5 after the other applied test methods have been described.

Table 5.2: Predictions for the $\frac{[H_2D^+]}{[H_3^+]}$ ratio for different experimental conditions based on Equation (5.28) in comparison to the measured values in continuous flow (Fig. 5.8) and freeze out mode (Fig. 5.10). The most probable values are printed in **bold face**.

	Continuous Flow				Freeze Out		
Parameters							
T_{gas}	12.8 K [a]	19 K [b]	**12.8 K** [a]	12.8 K [a]	12.8 K [a]	12.8 K [a]	**12.8 K** [a]
$\frac{[HD]}{[H_2]} \times 10^4$	3 [c]	3 [c]	**0.0201** [d]	0.0121 [e]	3 [c]	3 [c]	**0.42** [d]
$\frac{[\text{o-}H_2]}{[H_2]} \times 10^4$	9.23 [f]	9.23 [f]	**14.8** [g]	9.23 [f]	22.30 [h]	15.80 [g]	**2.78** [g]
Results for $\frac{H_2D^+}{H_3^+}$							
Prediction	2.2	1.5	**0.0090**	0.0090	0.87	1.25	**1.25**
Measurement	**0.0090 $\pm$ 0.0003**				**1.25 $\pm$ 0.10**		

[a] nominal trap temperature

[b] corrected trap temperature (see [7])

[c] natural HD amount

[d] corrected HD amount (see Fig. 5.9)

[e] HD amount that best fits the measured $\frac{H_2D^+}{H_3^+}$ ratio

[f] thermal equilibrium ratio of $\frac{[\text{o-}H_2]}{[H_2]}$ at 18.8 K

[g] ratio of $\frac{[\text{o-}H_2]}{[H_2]}$ that best fits the measured $\frac{H_2D^+}{H_3^+}$ ratio

[h] thermal equilibrium ratio of $\frac{[\text{o-}H_2]}{[H_2]}$ at 20.8 K

5.3 Testing the Purity of Para Hydrogen with Raman Spectroscopy

Raman Spectroscopy is a direct test mechanism for the ortho-to-para ratio of hydrogen and it is independent of the HD to H_2 ratio. By comparing the intensities of rotational transitions starting from either ortho or para levels, the ortho-to-para ratio can be derived. This method requires higher amounts of para hydrogen than the test reactions in the ion trap. Therefore, the para hydrogen generator was modified in this work as explained in Section 5.1.2. Now it allows the production of para hydrogen amounts that are sufficient for spectroscopic investigations. Parallel to the modifications on the para generator, a Raman spectrometer for solid state samples was modified by Juliane Gerke [20], to enable Raman spectroscopy of gaseous samples.

5.3.1 Raman Spectrometer

Raman spectroscopy is based on the inelastic scattering of light (i.e. the molecule absorbes a photon and immediately emits a photon of a shifted frequency. It is a method to obtain rotational and rovibrational spectra of molecules without permanent dipole moment, such as H_2. This section will give a brief introduction into the spectrometer used. More details on the principle and theory behind Raman spectroscopy can be found in [48]. Details on the experimental setup used in this work are presented in [20].

The Raman spectrometer used in this work was originally set up by the II. Physical Institute to investigate solid state samples. We were allowed to modify it in order to investigate gaseous samples as well. This modifications were part of the master's thesis of Juliane Gerke [20].

Figure 5.12 shows a schematic and Figure 5.13 shows two photographs of the modified Raman spectrometer. A laser beam is reflected by a set of mirrors to enter the gas cell from top to bottom. A small fraction of the laser light is Raman scattered by the molecules in the cell. In order to increase the amount of scattered light, a planar mirror below the cell reflects the beam for a second transition. The light that is scattered in the direction of the detection system first passes through a system of focusing lenses and an edge filter. Then the light passes through a slit and is reflected onto a CCD camera by a system of mirrors and gratings, which project different frequencies on different pixels of the CCD. The edge filter mentioned before is needed to filter out the elastically scattered light, which has much too high intensities for the CCD. By having the laser beam parallel to the slit (top to bottom), the amount of light reaching the CCD is maximized. This is necessary due to the low probability of Raman scattering.

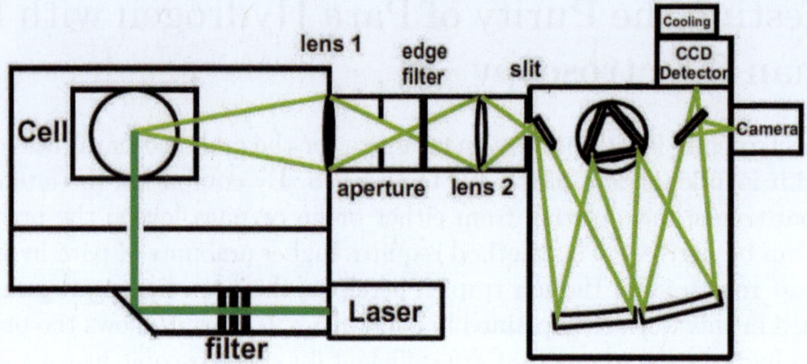

Figure 5.12: Schematic of the Raman spectrometer (view from the top). The laser beam *(dark green)* passes the glass cell from top to bottom. The light is scattered elastically and inelastically in all directions. Part of the scattered light *(light green)* enters the detection optics (for details, see text). From [20].

The low probability of Raman scattering sometimes hampered the use of Raman spectroscopy as diagnostic tool for the purity of para hydrogen. For pressures below ≈ 300 mbar, the intensity of the Raman scattered light becomes too small compared to the noise of the spectrometer.

In this work we follow the convention of using the Raman shift in units of rel cm^{-1} instead of the frequency for Raman measurements. It is given by the initial wavelength $\lambda_i = (532.2 \pm 0.3)$ nm of the laser and the Raman scattered wavelength λ_f as:

$$\Delta\tilde{\nu} = \left(\frac{1}{\lambda_i} - \frac{1}{\lambda_f}\right) \tag{5.31}$$

5.3.2 Raman Measurements

Figure 5.14 shows a Raman spectrum of normal hydrogen, observed with the modified Raman spectrometer described above. As explained in [20], the line intensities are given not only by the thermal energy distribution at room temperature, but also by spectrometer properties. In this work the first two lines are considered: the p-H$_2$ transition $J_{0\rightarrow 2}$ at (358.527 ± 0.014) rel cm^{-1}, and the o-H$_2$ transition $J_{1\rightarrow 3}$ at (590.337 ± 0.005) rel cm^{-1}. At room temperature, the equilibrium ratio $\frac{[\text{o-H}_2]}{[\text{p-H}_2]}$ is $\frac{3}{1}$. The influence on line intensities, that is due to thermal

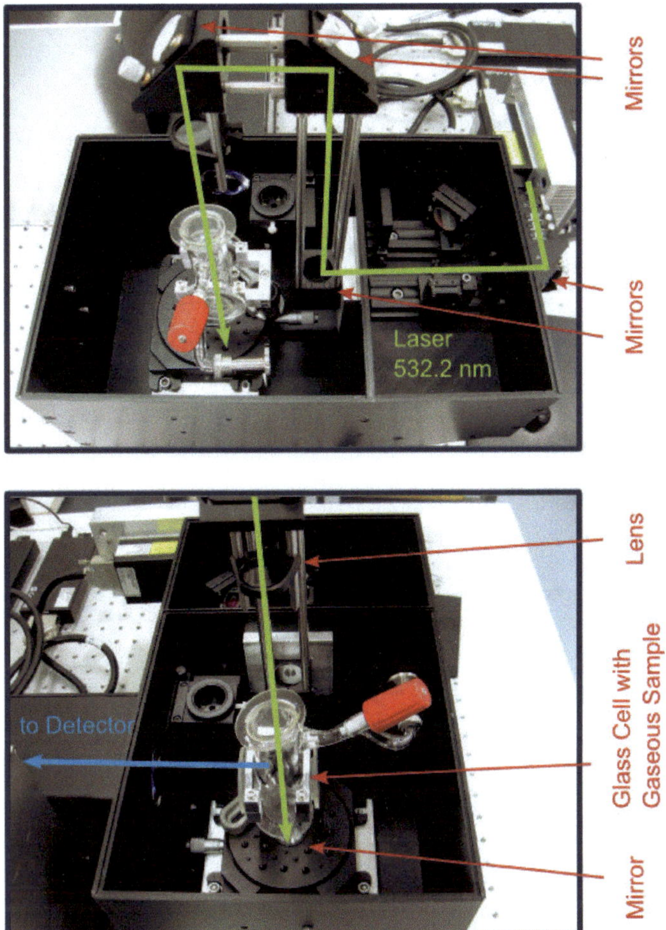

Figure 5.13: Photographs of the Raman spectrometer.
Top: View from the front. The laser beam is reflected by several mirrors to pass the glass cell containing the gas sample from top to bottom.
Bottom: View from the side. Before entering the cell, the laser beam is focused by a lens. Below the cell, a planar mirror reflects the beam back to get a second transition through the cell. The light scattered to the left enters the optics leading to the detector. More details can be found in [20].

level population or to spectrometer properties, is considered to be the same for all measurements. Thus, a spectrum of n-H_2 is used as reference for the calculation of line intensities of all p-H_2 samples. The ratio $\frac{[\text{o-}H_2]}{[\text{p-}H_2]}$ for any hydrogen sample can then be derived by comparison with the intensities I in n-H_2 as

$$\frac{[\text{o-}H_2]}{[\text{p-}H_2]}\bigg|_{\text{sample}} = \frac{I_o}{I_p}\bigg|_{\text{sample}} \cdot 3 \cdot \frac{I_p}{I_o}\bigg|_{\text{n-}H_2} . \tag{5.32}$$

A sample of p-H_2 (750 mbar) was produced at a maximal evaporation temperature of 20 K, to investigate the purity of the p-H_2 and the back conversion rate in the glass cell used for the Raman spectrometer.

Figure 5.15 shows the Raman spectrum of this sample. The measurement was taken over the period of 90 to 120 minutes after the production of the p-H_2 sample. As there is no ortho line visible, a Gaussian line profile with the same line width as the para lines was plotted at the position of the $J_{1\to3}$ line. The height of this line is the maximum intensity, that could be hidden in the noise of the spectrum. Thus, it is considered to be the uncertainty of the o-H_2 fraction, which is zero at that time.

Figure 5.16 shows the Raman spectrum of the same sample at a later time (six days after production). Here, the ortho line is clearly visible. A series of measurements was taken over a period of six days after production of the p-H_2. From each measurement, the $\frac{[\text{o-}H_2]}{[\text{p-}H_2]}$ fraction was determined using Eq. (5.32) which gives the ortho fraction as

$$\frac{[\text{o-}H_2]}{[H_2]} = \frac{[\text{o-}H_2]}{[\text{p-}H_2] + [\text{o-}H_2]}$$

$$= \left(\frac{[\text{p-}H_2]}{[\text{o-}H_2]} + 1\right)^{-1} . \tag{5.33}$$

The time evolution of the ortho fraction is supposed to follow

$$\frac{[\text{o-}H_2]}{[H_2]}(t) = \frac{3}{4} - \left(\frac{3}{4} - \frac{[\text{o-}H_2]}{[H_2]}\bigg|_{t=0}\right) e^{-\frac{t}{\tau}} \tag{5.34}$$

$$\approx f_0 + \left(\frac{3}{4} - f_0\right) \cdot \frac{t}{\tau} , \tag{5.35}$$

where (5.35) is the first term of a Taylor expansion for short times $t \ll \tau$ of (5.34) and $f_0 = \frac{[\text{o-}H_2]}{[H_2]}\big|_{t=0}$. The results shown in Figure 5.17 confirm, that the linear approximation is valid here. They lead to a time constant τ and an ortho fraction at $t = 0$ of:

$$\tau = (69 \pm 5) \text{ days}$$

$$\frac{[\text{o-}H_2]}{[H_2]}\bigg|_{t=0} = -0.009 \pm 0.004 \tag{5.36}$$

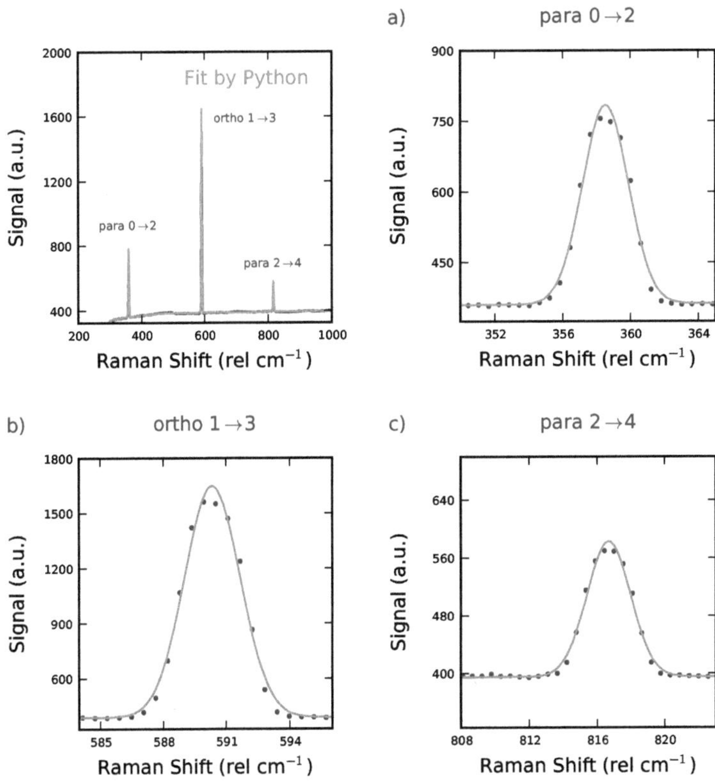

Figure 5.14: Raman spectrum of normal hydrogen (by Juliane Gerke [20]). Three rotational transitions of hydrogen are covered in the frequency range: $J_{0\to2}$, $J_{1\to3}$, and $J_{2\to4}$. To compare the intensities of these lines to the spectra of para hydrogen, the lines were fitted with Gaussian profiles in `python` [49]. Those fits are shown in the enlarged views *a)*, *b)*, *c)*.

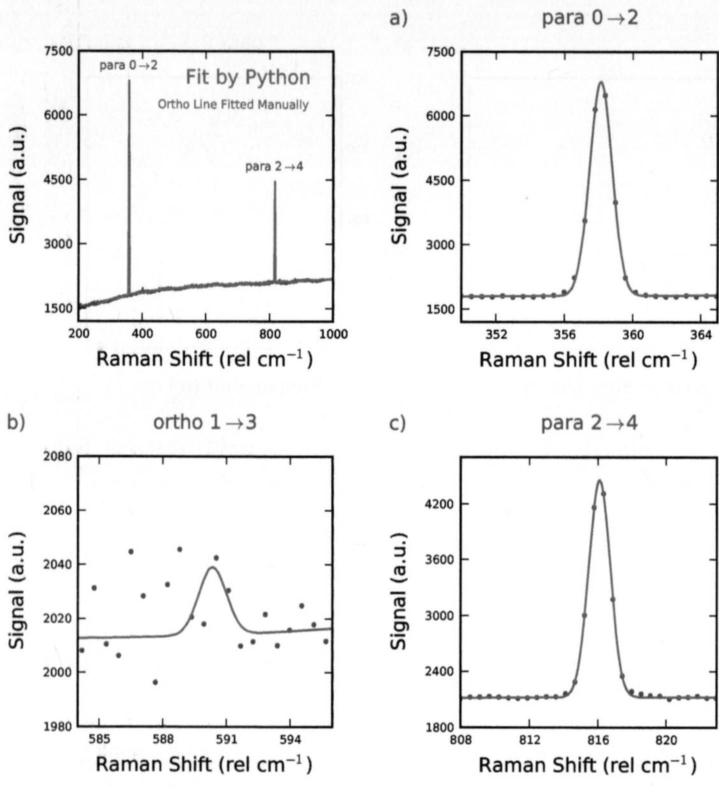

Figure 5.15: Raman spectrum of fresh para hydrogen. The same frequency range is covered as in Fig. 5.14. The o-H_2 line at 590.337 rel cm^{-1} is not visible in this measurement, as expected for pure p-H_2. To determine a detection limit for the spectrometer, a Gaussian profile was plotted at the expected position with the same width as the p-H_2 lines (see enlarged view *b)*). Comparison of the intensities of ortho and para lines (see Eq. (5.33)) reveals that the o-H_2 fraction in this sample was below 0.5 %.

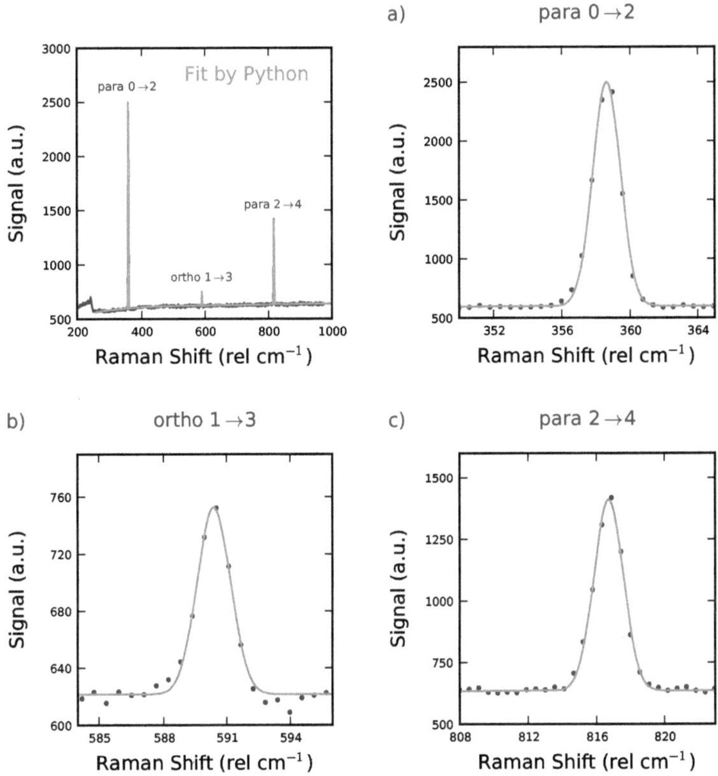

Figure 5.16: Raman spectrum of the same sample as in Figure 5.15 after six days storage in the glass cell of the Raman spectrometer. Due to the back conversion processes on the surfaces of the cell, the ortho fraction in the para hydrogen sample increases with time. After six days, the ortho line is clearly visible (see enlarged view *b)*).

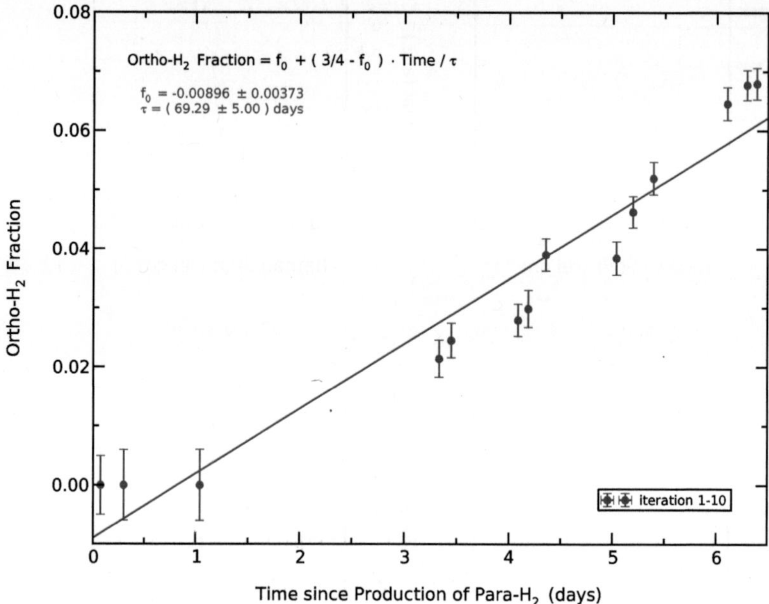

Figure 5.17: Time evolution of the o-H$_2$ fraction in a p-H$_2$ sample stored in the glass cell of the Raman spectrometer. The gas sample remained in the glass cell for six days in total. During this time, several Raman spectra were measured to derive the ortho fraction by Eq. (5.33). Each measurement is the average of ten iterations.

It is an open question, how to explain the negative value for the ortho fraction at $t = 0$, which was already observed before [20]. So far, the result for the purity of fresh p-H_2 is, that the o-H_2 fraction is below the Raman spectrometer's detection limit of 0.5 %. The thermal equilibrium at the highest evaporation temperature of 20 K for this sample is 0.16 % (see Table 5.1). However, the time constant for the back conversion is large enough, that for further measurements with the Raman spectrometer, the errors caused by back conversion in the glass cell can well be neglected.

The measurements presented in Figure 5.17 show rather large fluctuations around the expected linear curve. In order to analyse whether they are caused by some spectrometer drifts between two measurements or by rather short time scale effects within the measurements, the following investigations were performed.

Each measurement consists of ten iterations, each of them 180 seconds long. For the results shown so far, all ten iterations were averaged. If there were drifts with influence on the intensities on short timescales, the first iterations would yield different results than the last iterations of a measurement. Therefore, the first five iterations of each measurement and the last five iterations were averaged separately (an example is shown in Appendix A.5). With these results, the time dependence of the o-H_2 fraction was plotted again (Fig. 5.18).

For most measurements deviations during the measurement time are visible. Although the results for the time evolution as in Eq. (5.35) match astonishingly well, these fluctuations require further investigation.

Comparison of the enlarged views of the $J_{2\rightarrow4}$-lines in Figures 5.15 and 5.16 also reveals slight frequency shifts. As the determination of the p-H_2 purity does not need precise line positions, these shifts can only be relevant, if they happen on time scales shorter than the measurement time of 30 minutes. In this case, they would influence the line profile and the observed intensities. For the measurements performed in this work, this seems to be not the case. However, in further investigations, some caution should be taken with this point as well.

Measurements on the time constant of the back conversion in the Teflon storage bottle as in [20] ($\tau \approx 1000$ days) were not yet repeated, due to the large fluctuations observed. For the measurement series shown in Figure 5.18, the results for the time constant seem to be reliable despite these fluctuations, but here twelve measurements have been recorded. For the back conversion in the Teflon bottle, the number of measurements is limited to five. After that, the remaining p-H_2 pressure in the bottle will no longer be sufficient for Raman spectroscopy. When the p-H_2 is evaporated not only to the Teflon bottle but simultaneously to the glass cell, another measurement can be gained. But this has to be done very carefully, to avoid destruction of the cell by pressures higher than 1 atm. However, the quality of the Raman measurements is not reliable enough to derive the time constant from such few measurements.

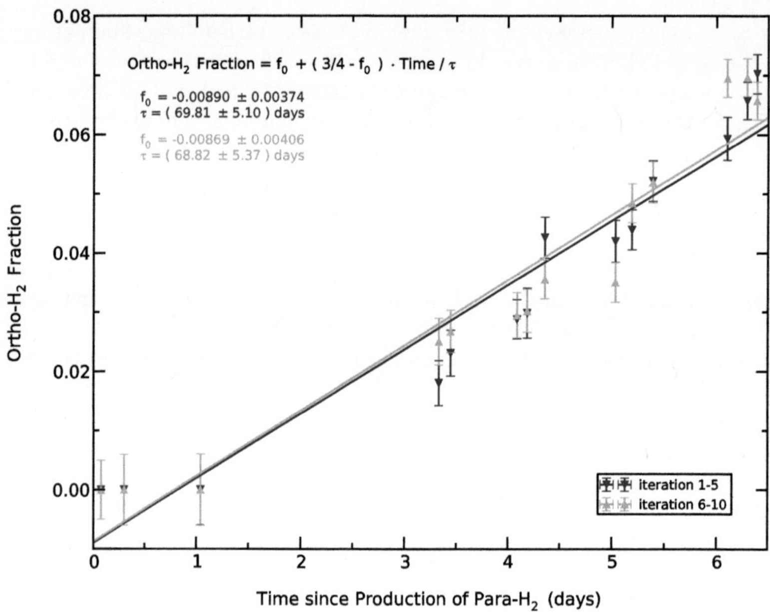

Figure 5.18: Time evolution of the o-H$_2$ fraction stored in a p-H$_2$ sample in the glass cell of the Raman spectrometer, including fluctuations within the measurements. For this plot, each measurement was split, i.e. the first five iterations ($\blacktriangledown$) and the last five iterations ($\blacktriangle$) were averaged and fitted as separate measurements. The fluctuations within the measurements are rather large. The two fits for the time dependence of the ortho fraction, however, match astonishingly well.

Chapter 6

$N^+ + H_2 \rightarrow NH^+ + H$
Dependence of the Reaction on the Internal Energies of the Reaction Partners

The reactions of NH_n^+ ions ($n = 0, \ldots, 3$) with hydrogen molecules are the basic steps in the formation of interstellar ammonia. The first reaction in this chain is known to be hindered by either an endothermicity or a barrier. It is not well enough understood yet, to decide, which of both mechanisms hinders the reaction and what energy is required to make the reaction possible. So far, it is well known that the difference in rotational energies between ortho and para hydrogen influences the reaction rate by several orders of magnitude at low temperatures (a few ten Kelvin). The first excited rotational level of H_2 yields an energy amount (173 K), that is in the same order of magnitude as the energy required to enable the reaction with N^+ (≈ 230 K). It is an open question, to what extent the energy of the three lowest fine-structure states of N^+ is able to enhance the reaction.

Due to its sensitivity on the ortho-to-para ratio of the reactant hydrogen, the reaction rate of an H_2 sample with N^+ will also serve as a precise test method for this ratio. This application requires a detailed understanding of the reaction.

6.1   The $N^+ + H_2$ Reaction System

The formation of interstellar ammonia starts with the following reaction chain:

$$N^+ + H_2 + E \underset{k_{b1}}{\overset{k_{f1}}{\rightleftharpoons}} NH^+ + H \tag{6.1}$$

$$NH^+ + H_2 \underset{k_{b2}}{\overset{k_{f2}}{\rightleftharpoons}} NH_2^+ + H + E \tag{6.2}$$

$$NH_2^+ + H_2 \underset{k_{b3}}{\overset{k_{f3}}{\rightleftharpoons}} NH_3^+ + H + E \tag{6.3}$$

$$NH_3^+ + H_2 + E \underset{k_{b4}}{\overset{k_{f4}}{\rightleftharpoons}} NH_4^+ + H \tag{6.4}$$

These reactions have been investigated by various methods in several groups over the last three decades. A nice summary of previous experiments is given in [50]. So far, reactions (6.2) and (6.3) are known to be exothermic, while (6.1) and (6.4) require energy input. The focus of this work is on reaction (6.1). The results for the required energy for this reaction vary in earlier publications. It is supposed to be around 230 K, but it is still unclear, whether it is due to an endothermicity or a barrier. For simplicity, in the following just the term endothermicity will be used.

We started our investigations of the reaction system (6.1) to (6.4) by observing the temporal evolution of all species NH_n^+ ($n = 0, \ldots, 4$) when N^+ and p-H_2 or n-H_2 were injected in the 22-pole trap. These measurements were performed at nominal temperatures in the range of 10 to 300 K. The ions were cooled down to the ambient temperature of the 22-pole trap by an intense helium pulse at the beginning of each experimental cycle.

Special care was taken to optimize all trapping potentials, so that the sum of all ions of different species stays constant with the storage time, i.e. all species are detected with the same efficiency. Figure 6.1 shows an example of such a measurement with p-H_2 at ≈ 11 K. The total number of ions in the trap is a constant and the temporal evolution can be described by numerical simulations of the forward reactions of (6.1) to (6.4), the back reactions can be neglected, since no H atoms are present in the 22-pole trap chamber (H_2 is flowing through the trap vacuum chamber continuously). From these results we can conclude, that no information on the forward reaction rate of (6.1) will be lost, when only the decrease of N^+ ions with the storage time is recorded. The time consumption of the measurements can be reduced by a factor of five, when only one product

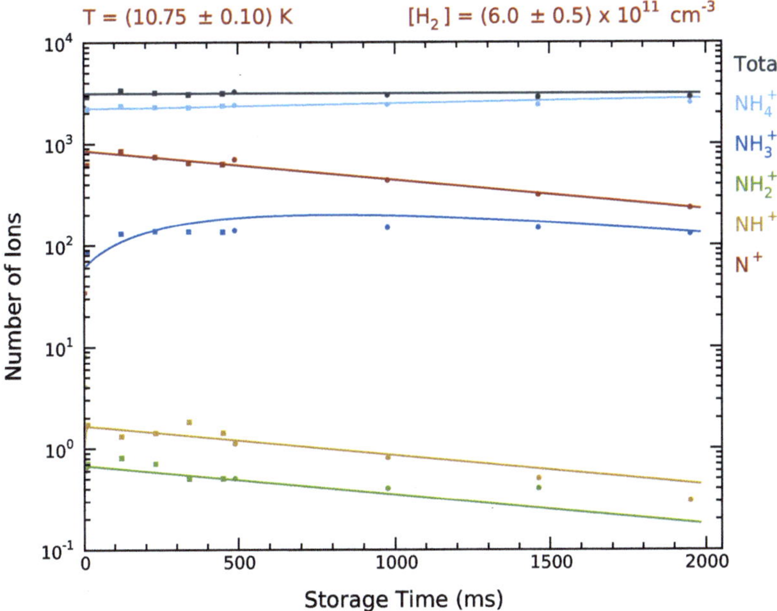

Figure 6.1: All reaction products of the N^+ + p-H_2 system at a nominal trap temperature of ≈ 11 K.

At the beginning of the project, several series of measurements were performed, counting all possible reaction products. The sum over all ion species is constant over the reaction time and the time evolution of each species can be reproduced by numerical simulation of the forward reactions of (6.1) to (6.4). This ensures, that for further measurements no important information is lost, when only the decrease of N^+ ions is recorded.

The following rate coefficients and ion numbers at $t = 0$ were used for the numerical simulation:

$N^+_0 = 850$, $k_{f1} = 1.1 \times 10^{-12} \frac{cm^3}{s}$, $NH^+_0 = 1$, $k_{f2} = 5.7 \times 10^{-10} \frac{cm^3}{s}$, $NH^+_{2\,0} = 1$, $k_{f3} = 1.4 \times 10^{-9} \frac{cm^3}{s}$, $NH^+_{3\,0} = 60$, $k_{f4} = 2.8 \times 10^{-12} \frac{cm^3}{s}$, $NH^+_{4\,0} = 2200$.

Although only N^+ was admitted to the 22-pole trap, the simulation has to start with nonzero numbers of the other ions as well. This is due to the reaction of hot ions in the first milliseconds of the storage time, while they were still relaxing in He-collisions. The quite high number of initial NH_4^+ ions might be due to impurities of n-H_2 in the He buffer gas. Those would rapidly convert N^+ ions injected to the trap to NH_4^+ via reactions (6.1) to (6.4). However, the derived rate coefficient for the formation of NH_4^+ from NH_3^+ is consistent with the results from [51].

For the measurements shown here, the ions were produced with an ionization energy of 80 eV. After entering the 22-pole trap, they were cooled down to the ambient temperature by a Helium pulse.

mass is counted instead of five. This reduces the temperature shifts during the measurement and thereby improves the quality of the measurements dramatically. Only temperatures below 30 K can be set as constant values by heating against the running cold head. Higher temperatures can only be reached by cooling down from room temperature or by heating up with deactivated cold head from cryogenic temperatures. Thus, it is preferable to record the temporal evolution in less than ten minutes, to keep temperature changes small.

6.2 Influence of the Temperature and the Ortho-to-Para Ratio of H_2 on the Reaction Rate

Reaction (6.1) is endothermic by $\approx$ 230 K. This energy can be provided by any form of kinetic energy of the reaction partners, i.e. translational energy from H_2 or N^+ or rotational energy from H_2. Vibrational excitation of H_2 ($\approx$ 1000 K) can be neglected at the experimental temperatures (10 to 300 K). N^+ does not have rotational or vibrational levels. It is still an open question, to what degree the fine-structure energy of N^+ is available to overcome the endothermicity of reaction (6.1) (see Section 6.3).

In the easiest case, the lowest possible level for the reacting H_2 molecule is its rotational ground state (p-H_2, $J = 0$). Then, the probability to gain the activation energy E_A required for the reaction with N^+ at a certain temperature T is given by the Boltzmann factor. Therefore, the temperature dependence of the rate coefficient can be described by an Arrhenius equation:

$$k_{f1}(T) \;=\; k_A \exp\left(-\frac{E_A}{k_B T}\right), \tag{6.5}$$

where k_A is the rate coefficient for $\frac{1}{T} = 0$ and k_B is the Boltzmann constant. From the rate coefficient k (in units of $\frac{cm^3}{s}$), we can derive the reaction rate K (in units of s^{-1}) for a number density $[H_2]$ (in units of cm^{-3}) of the reactant hydrogen as

$$K \;=\; k \cdot [H_2]. \tag{6.6}$$

For ortho hydrogen, the lowest possible state is the first excited rotational state ($J = 1$ at 173 K). Therefore, the energy that has to be gained from the thermal energy is reduced. The best guess would be, that $\frac{E_A}{k_B}$ has to be reduced by 173 K for o-H_2.

From these considerations, we can expect a very steep temperature dependence for the rate coefficient in reactions of N^+ with p-H_2 and a comparatively flat dependence for reactions with o-H_2. The temperature dependent rate coefficient of any hydrogen sample that contains an ortho fraction f is then given by

$$k_{f1}(T) \quad = \quad f k_o(T) + (1 - f) \, k_p(T) \,, \tag{6.7}$$

$$f \quad = \quad \frac{[\text{o-}H_2]}{[H_2]} \,.$$

Figures 6.2 and 6.3 show measurements of the decrease of N^+ ions with time in
reaction with n-H_2 (75 % o-H_2, 25 % p-H_2) and p-H_2 at different temperatures.

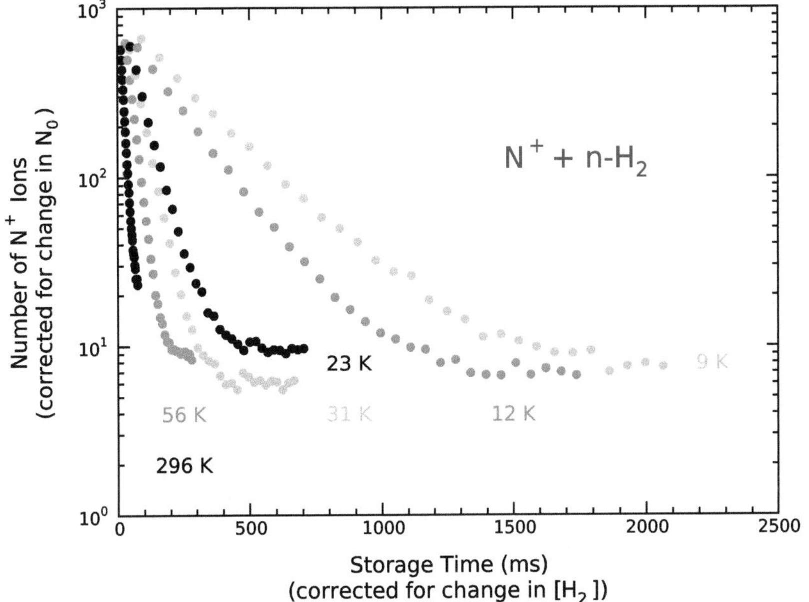

Figure 6.2: Decrease of N^+ ions with the storage time in reaction with n-H_2 at
different temperatures.
For better comparability, each measurement was corrected for experimental drifts
(H_2 number density and number of N^+ ions at $t = 0$). The reaction rate decreases
towards lower temperatures.
For the measurements shown here, the ions were produced with an ionization
energy of 60 eV. After entering the 22-pole trap, they were cooled down to the
ambient temperature by a Helium pulse.
The measurements do not follow a mono exponential decay, but evolve towards
a constant background. This effect will be discussed later.

They qualitatively confirm the expectations. In both cases, the reaction slows down towards lower temperatures, but the effect is much stronger for p-H_2.

A plot of the logarithm of the rate coefficients versus $\frac{1}{T}$ (as in Figure 6.6) would then result in a steep straight line (p-H_2) bending into a flatter straight line (o-H_2) towards low temperatures. At which temperature the bend occurs, i.e. how high the flat straight line is, will be given by the ortho-to-para ratio of the reactant hydrogen. This forecast cannot be confirmed completely. Figure 6.6 reveals, that the temperature dependence is more complicated than considered so far. The bending of the curves from a steep slope into a flatter slope is visible,

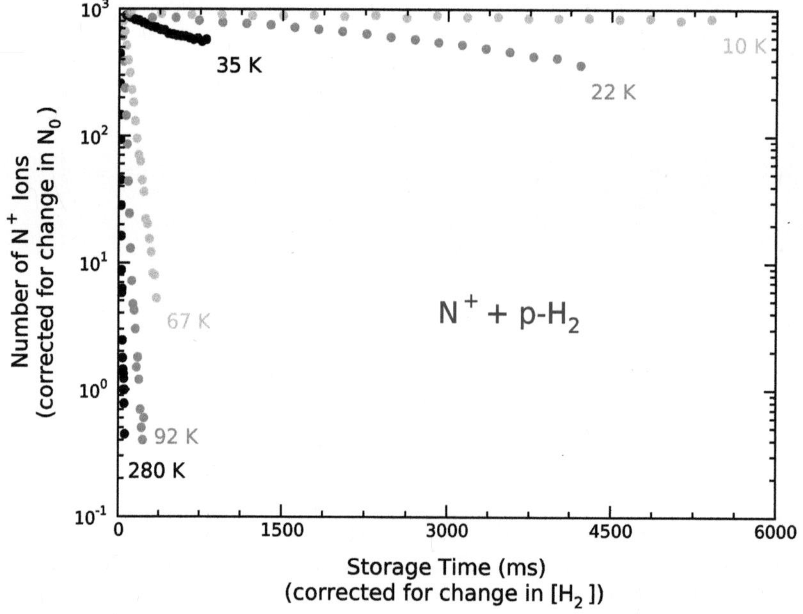

Figure 6.3: Decrease of N^+ ions with the storage time in reaction with p-H_2 at different temperatures.

For better comparability, each measurement was corrected for experimental drifts (H_2 number density and number of N^+ ions at $t = 0$). The reaction rate decreases towards lower temperatures. This effect is much stronger than for the measurements with n-H_2 shown in Fig. 6.2.

For the measurements shown here, the ions were produced with an ionization energy of 60 eV. After entering the 22-pole trap, they were cooled down to the ambient temperature by a Helium pulse.

but it obviously cannot be described by just two straight lines. Hence, the energy of the fine-structure levels of N$^+$ needs to be taken into account.

6.3 Influence of the Fine-Structure of N$^+$

6.3.1 Fine-Structure of N$^+$

The electronic ground state of N$^+$ is $1s^2 2s^2 2p^2$. Spin-orbit coupling of the electrons causes the splitting of internal states that is called fine-structure. Here we will give a brief summary of the consequences of spin-orbit coupling for the internal states of N$^+$. More details on this effect can be found in e.g. [48] and [52].

The two closed s-shells in N$^+$ do not have a total angular momentum or total electron spin. Thus, they can be neglected for the fine-structure. The two electrons in the p-shell are equivalent. Their orbital angular momentum $\vec{L}$ couples with their spin $\vec{S}$. Since we have two particles with $l = 1$ and $s = \frac{1}{2}$, the total orbital angular momentum can take the values $L = 0, 1, 2$ and the total spin can be $S = 0, 1$. There are six possible combinations $^{2S+1}L_J$, $\vec{J} = \vec{L} + \vec{S}$ (compare Figure 6.4): ^1S_J, ^1P_J, ^1D_J, ^3S_J, ^3P_J, ^3D_J. Three of them are forbidden by the Pauli exclusion principle, namely ^1P_J, ^3S_J, ^3D_J. For a given L and S, J can take values between $L + S$ and $|L - S|$. Thus, the resulting fine-structure states of N$^+$ are: $^3P_0$, $^3P_1$, $^3P_2$, $^1D_2$, $^1S_0$ (in the order of increasing energy). Each of these states is $2J + 1$ times degenerate.

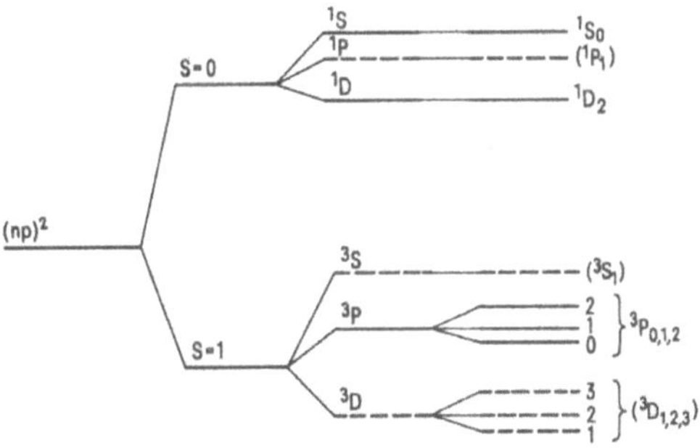

Figure 6.4: Spin-orbit coupling in N$^+$. For two equivalent electrons in a p-orbital, the total orbital angular momentum couples with the total electron spin to the shown states. The dashed lines are forbidden by the Pauli exclusion principle. From [52].

Table 6.1: Degeneracies g_i and energies E_i for the three lowest fine structure states of N^+, from [50].

^3P_i	g_i	$\frac{E_i}{k_B}$ (K)
$^3P_0$	1	0
$^3P_1$	3	71
$^3P_2$	5	188

6.3.2   Ratecoefficients Including the Fine-Structure of N^+

The three lowest fine-structure levels of N^+ are given in Table 6.1. They contain an amount of energy, that is of the same order of magnitude as the endothermicity of the reaction of N^+ with H_2. This fine-structure energy might enhance the reaction rate in the same way the rotational energy of o-H_2 does.

The latest results for the influence of the fine-structure population of N^+ on the reaction rate have been published by Zymak et al [50]. In this section we will compare our results to theirs.

The approach by Zymak et al [50] is, that not only each nuclear spin configuration of H_2 but also each of the three lowest fine-structure states of N^+ requires its own state-specific rate coefficient. For symmetry reasons, they consider the $^3P_2$ state to be non reactive. This leaves two fine-structure states and two nuclear spin configurations that need to be taken into account. Thus, we end up with four terms in the description of the rate coefficient's temperature dependence:

$$
\begin{aligned}
k_{f1}(T) \;=\; & f \quad (\zeta_0(T)k_{1,0}(T) + \zeta_1(T)k_{1,1}(T)) \\
& + \;(1-f)\,(\zeta_0(T)k_{0,0}(T) + \zeta_1(T)k_{0,1}(T)) \qquad (6.8)
\end{aligned}
$$

The temperature dependence of each state-specific rate coefficient is given by the Arrhenius equation:

$$
k_{j,i}(T) \;=\; k_{j,i}^A \exp\left(-\frac{E_{j,i}^A}{k_B T}\right) \qquad (6.9)
$$

The factors $k_{j,i}^A$ describe the reaction rate for high temperatures and $E_{j,i}^A$ are the activation energies for the reaction of p-H_2 ($j = 0$) or o-H_2 ($j = 1$) with the two lowest fine structure states of N^+, $^3P_0$ ($i = 0$) or $^3P_1$ ($i = 1$). The values derived by Zymak et al are given in Table 6.2. The ζ_i in Eq. (6.8) give the thermal fine-structure population of the state i:

$$
\zeta_i(T) \;=\; \frac{g_i}{Z(T)} \exp\left(-\frac{E_i}{k_B T}\right) \qquad (6.10)
$$

Table 6.2: State specific rate coefficients $k_{j,i}^A$ and activation energies $E_{j,i}^A$ for the reaction of p-H_2 ($j = 0$) or o-H_2 ($j = 1$) with the two lowest fine structure states of N^+, $^3P_0$ ($i = 0$) and $^3P_1$ ($i = 1$), from [50].

$k_{j,i}$	$k_{j,i}^A \left(10^{-10} \frac{\text{cm}^3}{\text{s}}\right)$	$\frac{E_{j,i}^A}{k_B}$ (K)
$k_{0,0}$	12	230
$k_{1,0}$	1.9	18
$k_{0,1}$	14	230
$k_{1,1}$	12	40

$Z(T)$ is the partition function:

$$Z(T) \;=\; \sum_{i=1}^{3} g_i \exp\left(-\frac{E_i}{k_B T}\right) \tag{6.11}$$

Table 6.1 gives the degeneracies g_i and the energies E_i of the three considered fine-structure states $^3P_0$, $^3P_1$, and $^3P_2$. The thermal population of these levels as derived from function (6.10) is shown in Figure 6.5. Here, we need to mention, that it is not yet clear, whether the fine-structure population of N^+ is relaxed by collisions with H_2 and He fast enough in the trap, to justify the assumption of a thermal population (compare Section 6.4.1). However, this is the best theory available so far.

6.3.3 First Experimental Results

Figure 6.6 compares three curves for $k_{f1}(T)$ measured in this work to the function (6.8) from [50]. In *dark red*, the results for normal hydrogen are shown. Our measured values lie slightly beneath the solid line of function (6.8) and they show a different curvature. For the low temperature region, they would better be fitted by the *light red* line, corresponding to an ortho fraction of 60 % instead of 75 %. This is a clear indication of systematic deviations between our results and those from [50], as the natural ortho fraction is beyond any doubt. In *green*, our best results for para hydrogen are shown together with the best fit of function (6.8). From this fit the ortho fraction in the p-H_2 sample would be derived to be ≈ 0.15 %. The *blue* line represents the lowest ortho fraction of 0.5 % that was observed by Zymak et al. Additionally, a measurement with a mixed sample of n-H_2 and p-H_2 is shown in *purple*. From the partial pressures of n-H_2 and p-H_2, an ortho fraction of (18.75 ± 1.55) % was calculated for this sample. The corresponding plot of function (6.8) is shown in *dark purple*, the best fitting function (corresponding to an ortho fraction of 12 %) is shown in *light purple*. Since the best measurements for n-H_2 and for the 18.75 % mixture lie below the corresponding functions, it is likely, that also the p-H_2 curve shows this deviation. Thus, the resulting ortho fraction of 0.15 % should be considered to have an uncertainty of 50 %:

$$\Rightarrow \quad \frac{[\text{o-}H_2]}{[H_2]} \;=\; (0.150 \pm 0.075)\ \% \tag{6.12}$$

The maximal evaporation temperature of this para hydrogen sample was 17.3 K. The ortho fraction in thermal equilibrium at this temperature is 0.04 %.
Figure 6.6 also shows the predicted lower detection limit for this kind of measurements *(yellow dashed lines)*. It is given by the reaction of N^+ ions with HD molecules, that are naturally admixed to the reactant H_2 gas:

$$N^+ + HD \;\leftrightharpoons\; NH^+ + D \tag{6.13}$$
$$\leftrightharpoons\; ND^+ + H \tag{6.14}$$

From the consideration of zero point vibrational energies, the formation of NH^+ is expected to be endothermic, the formation of ND^+ to be exothermic. Both

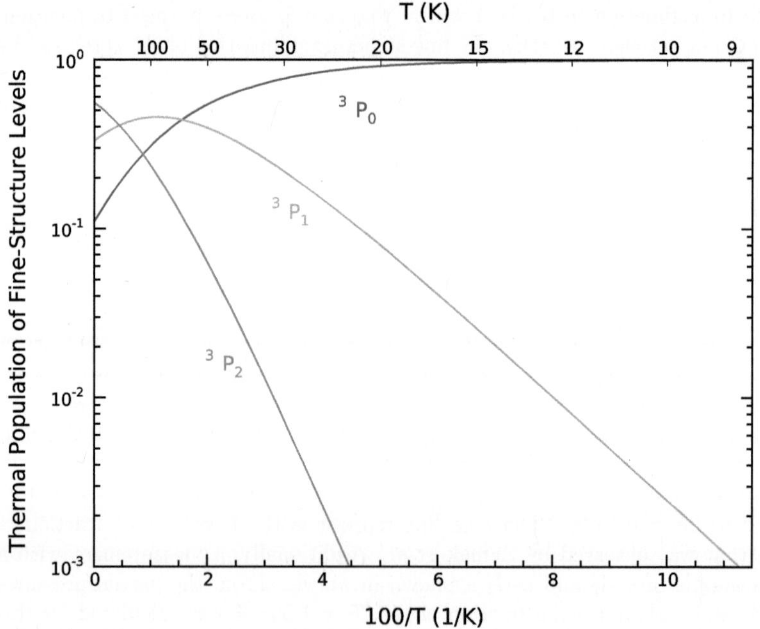

Figure 6.5:
Thermal population of the fine-structure levels $^3P_0$, $^3P_1$, and $^3P_2$ of N^+.

may be hindered by barriers, but accurate values are lacking. However, the overall rate for the destruction of N^+ by HD might be expected to be comparable to the rate for reactions of N^+ with o-H_2. The *yellow* curve in Figure 6.6 represents the low temperature behaviour of the rate coefficient observed recently by Gerlich [53], scaled to the highest possible ratio of HD-to-H_2 of 3.8×10^{-4}. This ratio is increased with respect to the natural HD amount of 3×10^{-4}, due to different pumping efficiencies for HD and H_2 [53]. As explained in Chapter 5, the HD amount in p-H_2 might be reduced by preferential freezing in the para generator. Thus, the plotted curve represents the worst case of a possible detection limit. For higher temperatures, the rate coefficients observed for the reaction

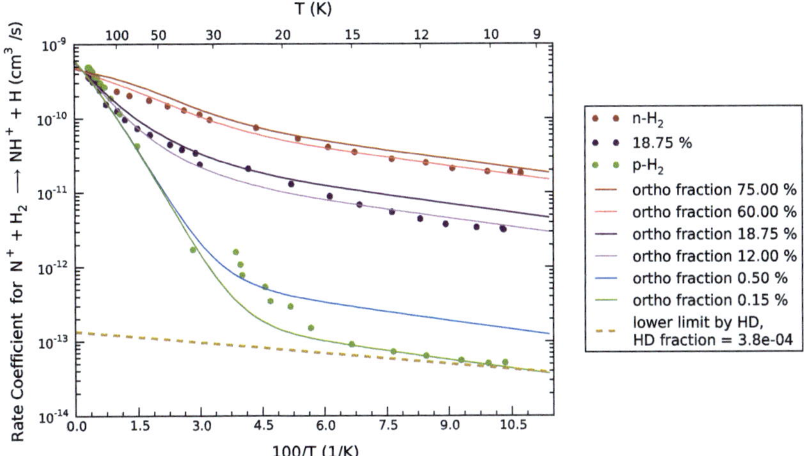

Figure 6.6: Temperature dependence of the rate coefficient for the reaction of N^+ ions with n-H_2, p-H_2 and a mixture containing 18.75 % o-H_2. The temperatures given here are the nominal trap temperatures.
Red circles: Measurement series with n-H_2. Ionization energy: 60 eV, Helium pulse for thermalization of the ions. (See also Fig. 6.2.)
Purple circles: Measurement series with a mixture containing (18.75 ± 1.55) % o-H_2. Ionization energy: 25.4 eV, Helium pulse for thermalization of the ions.
Green circles: Measurement series with p-H_2. Ionization energy: 60 eV, Helium pulse for thermalization of the ions. (See also Fig. 6.3.)
Solid lines: Function (6.8) from [50] for different o-H_2 fractions (see legend).
Blue line: Lowest rate coefficients observed by Zymak et. al. [50].
Dashed line: Lower detection limit by reactions with the natural amount of HD in the reactant gas H_2. $k(T) = 3.5 \times 10^{-10} e^{-\frac{11\,K}{T}} \, \frac{cm^3}{s}$, from [53].

of N^+ with HD deviate from the mono exponential line in a comparable way to the deviations observed for ortho- and para-H_2, strengthening the theory of the influence of the N^+ fine-structure energy on the reaction. Measurements of the $N^+ +$ HD reaction ratecoefficient are shown in the Appendix A.6.1.

6.4   Influence of Different Experimental Parameters

Figure 6.6 shows the best three examples of a large number of measurement series, taken to understand reaction (6.1). It turned out, that it is very hard to get reproducible results for the temperature dependence of the rate coefficient. Almost every experimental parameter seems to influence the observed speed of the reaction. The observed effects will be discussed in this section. Three correlations were investigated in more detail: the influence of the buffer gas density (Section 6.4.1), the influence of the ionization energy (Section 6.4.2), and the influence of the amplitude of the trapping radio frequency (RF, Section 6.4.3). Besides these correlations, the following effects need to be considered:

- **Cycle Time of the Experiment:** At the beginning of the experiments, the cycle time (i.e. the time between two extractions of ions from the source) was kept as short as possible, to reduce the time consumption of the measurements. The reaction is faster at higher temperatures than at low ones. Thus, the maximal storage time that has to be investigated is smaller for higher temperatures. In principle, a shorter storage time would allow to choose shorter cycle times. It turned out, that variations of the cycle time influence the observed rate coefficients. This is most likely due to relaxation of the N^+ ions, that gain plenty of energy in the ionization process in the source. When the ions are extracted from the source after a shorter cycle time, they experienced less relaxing collisions with the source gases. For high ionization energies (60 eV) an increase of the reaction rate towards shorter cycle times was observed, but no systematic measurements were performed on this effect. However, to get comparable results, all measurements have to be performed at the same cycle time.

- **N_2 Pressure in the Source:** For some measurements, there seemed to be an influence of the source pressure on the observed rate coefficients. Such effects could be explained by the relaxation processes after ionization mentioned above. However, they were not reproducible enough to derive any systematic relation.

- **He Pressure in the Source:** Adding He to the source gas dramatically increases the amount of produced N^+ ions and improves the ability to trap them at high temperatures. Here again the relaxation after ionization is the best explanation. No measurements were performed on the quantitative description of this effect.

- **Electric Potentials of the Ion Trap Apparatus:** When the ionization energy is varied, it is usually necessary to modify some other trapping or guiding potentials as well. Unknown effects of these potentials on the very sensitive reaction system might explain some of the problems concerning reproducibility of the measurement results.

- **Frequency of the Trapping RF:** As the parts of the 22-pole trap shrink a bit at low temperatures, the resonance frequency of the device changes with temperature. When the trap is operated at other frequencies than the resonance frequency, the amplitude of the trapping field decreases. As the amplitude has to be optimized for the experiments on N^+, the frequency of the RF generator has to be optimized for each experimental temperature.

- **Number Density of the Reactant Hydrogen:** At the beginning of the experiments, there seemed to be an effect of the number density of H_2 in the 22-pole trap chamber on the observed rate coefficients. Systematic measurements on this effect revealed, that it can be reduced enough to be negligible in the pressure range usually chosen for the experiments. This can be achieved by operating the ionization pressure gauge at a current of 1 mA instead of the usually used 0.1 mA. It turned out that 0.1 mA was not sufficient to measure low number densities correctly, resulting in wrong rate coefficients.

- **Time since the opening of the p-H_2 leakage valve:** Several series of measurements were performed (at room temperature and at $\approx$ 10 K) to check for experimental drifts, happening when no experimental parameter is modified intentionally. The observed rate coefficients showed variation up to 20 % at room temperature and up to 50 % at 10 K. In most, but not all cases these variations appeared to be anti proportional to the variations of the number density of H_2. For long operation times (several hours), the leakage valves tend to close somewhat. This is most likely due to the following processes: The expanding gas cools down the surfaces of the valve. At some point, they are cold enough, that molecules (perhaps water impurities) start to freeze out. By this freezing, the leak starts to close, reducing the pressure in the ion trap. This effect can be prevented by higher gas fluxes through the valve. Unfortunately, this increases the reaction rate so much, that measurements at high temperatures become impossible. How the change of the number density influences the observed reaction rate coefficients is still an open question. For low temperatures, there might be some influence by HD freeze out effects (see Chapter 5). The longer the trap is operated at 10 K with H_2 flowing through the apparatus, the more HD is frozen on the surfaces. This increases the amount of HD in the H_2 gas, leading to higher reaction rates. This effect cannot happen at high temperatures, but it might explain, why the time-dependence of the observed rate coefficients is stronger at low temperatures.

6.4.1   Influence of the Buffer Gas Density - Steps towards State-Specific Rate-Coefficients

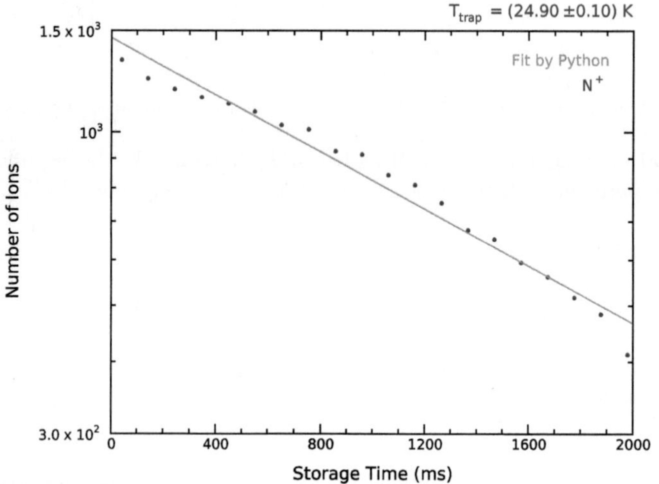

Figure 6.7: Example for a measurement showing a deviation from the mono exponential reaction *(solid line)*. Most measurements in the temperature range between $\approx$ 20 and 50 K show this behavior. For the measurement shown here, He was admitted to the 22-pole trap chamber via the pulsed valve at the beginning of each experimental cycle. The ionization energy in the ion source was 60 eV. The following rate coefficient and ion number at $t = 0$ were used for the numerical simulation:

$N^+_0 = 1457$, $k_{eff} = 7.7 \times 10^{-13}\ \frac{cm^3}{s}$ (effective reaction rate coefficient of N^+ (all levels) with H_2 (both ortho and para)).

The observed rate coefficients of the forward reaction of (6.1) showed dependencies on many experimental parameters. In spite of choosing the best possible experimental conditions, the measurement results were not always reproducible. For these measurements, the only parameter that was not considered to have an influence and was therefore changed without monitoring its value, was the intensity of the helium pulse.

The observed variations might be explained by insufficient cooling of the fine-structure levels of N^+. Another hint towards this theory is the shape of some of the observed time evolutions in the reaction system. For nominal temperatures around 30 K, almost all N^+ curves are convex (see e.g. Figure 6.7). This effect is exactly in the temperature region, where the observed temperature dependence of the rate coefficient k_{f1} deviates from the predictions of any theory (the rate coefficients are to large). This could be explained, when two levels with very

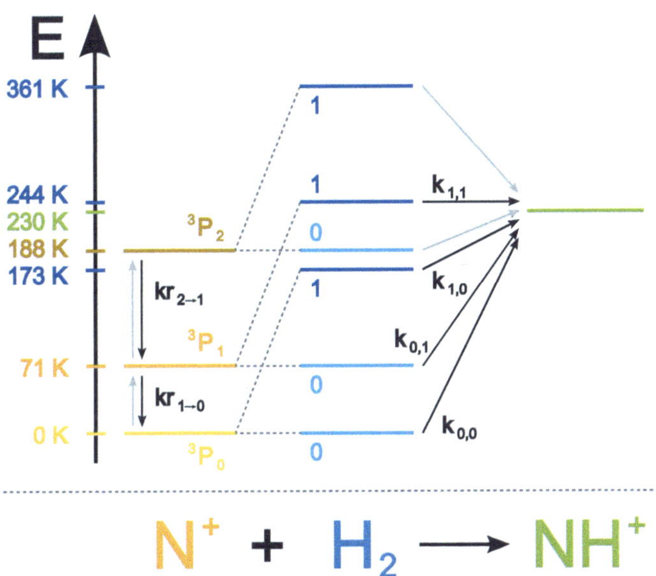

Figure 6.8: Scheme of state-specific reaction and cooling rate coefficients as introduced in [50].
The two lowest fine-structure states ($^3P_0$, $^3P_1$) of N^+ *(yellow)* can react with the two lowest rotational levels ($J = 0, 1$) of H_2 *(blue)* to form NH^+ *(green)*. Each combination of levels has another rate coefficient $k_{j,i}$ for the reaction. The fine-structure level ($^3P_2$) is considered to be non reactive. Additionally, relaxation from $^3P_2$ to $^3P_1$ ($kr_{2\to1}$) and from $^3P_1$ to $^3P_0$ ($kr_{1\to0}$) is allowed (via collisions with H_2 or He). The *light grey* arrows mark those transitions that are neglected in this model.

different rate coefficients were involved in the reaction. If the cooling rate from the slow to the fast reacting level is comparable to the reaction rate of the slow level, the reaction accelerates with the storage time. This scheme of rate coefficients is shown in Figure 6.8. A fit of this model to the measurement presented in Figure 6.7 is shown in Figure 6.9

The first fits that included state specific reaction and fine-structure relaxation rates looked very promising, since they were able to reproduce the observed curves almost perfectly. However, the model used here becomes unphysical at temperatures higher than 30 K, where a significant amount of N$^+$ ions is expected to be in one of the two excited fine-structure states in thermal equilibrium. It makes

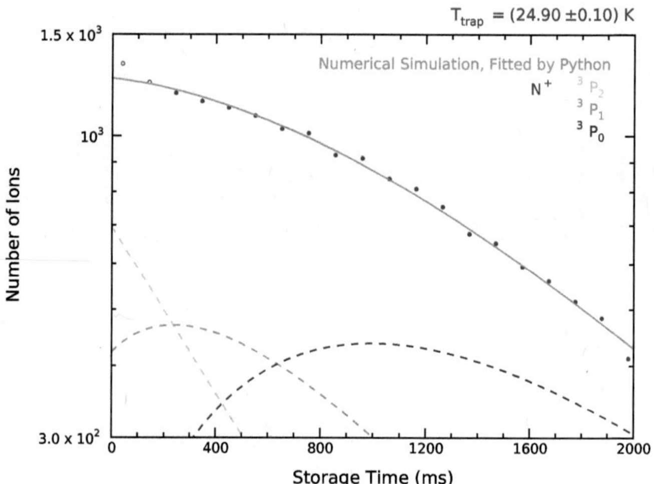

Figure 6.9: The same measurement as in Fig. 6.7, fitted with state specific rate coefficients. The *solid* curve is the result of a numerical simulation, based on the reaction and relaxation rates shown in Fig. 6.8. Also shown is the time evolution of the population of each fine-structure state, resulting from the simulation. The first two measurement points *(empty circles)* have been excluded from the fit, because at those times, translational cooling of the ions might be still in progress. The following rate coefficients and ion numbers at $t = 0$ were used for the numerical simulation:
N$^+_0$ = 1260, $k_{1,eff} = 5.5 \times 10^{-19} \frac{cm^3}{s}$ (effective reaction rate coefficient of the level ^{3}P$_1$ with H$_2$ (both ortho and para)), $k_{0,eff} = 1.6 \times 10^{-12} \frac{cm^3}{s}$ (effective reaction rate coefficient of the level ^{3}P$_0$ with H$_2$), $k_{relax} = 2.2 \times 10^{-12} \frac{cm^3}{s}$ (relaxation rate coefficient from ^{3}P$_1$ to ^{3}P$_0$ in collisions with H$_2$).

no sense, that inelastic collisions with neutral gas should result in cooling all the ions to the ground state at such temperatures. Therefore, the model was extended by rate coefficients that bring N^+ ions to the next higher fine-structure level. They were connected to the ones for fine-structure relaxation by Boltzmann factors, so that the fine-structure population of the N^+ sample would develop towards the thermal equilibrium, if reactions were neglected (compare Eq. (8.13) in Chapter 8.2.2).

Fits with the extended model do not result in bent curves for temperatures higher than 30 K. They rather form of the combination of two linear curves (on the logarithmic scale). For early times, the slope is given by the fastest rate in the system, for late times by the slowest. Hence, another explanation is needed for the observed convex time evolutions.

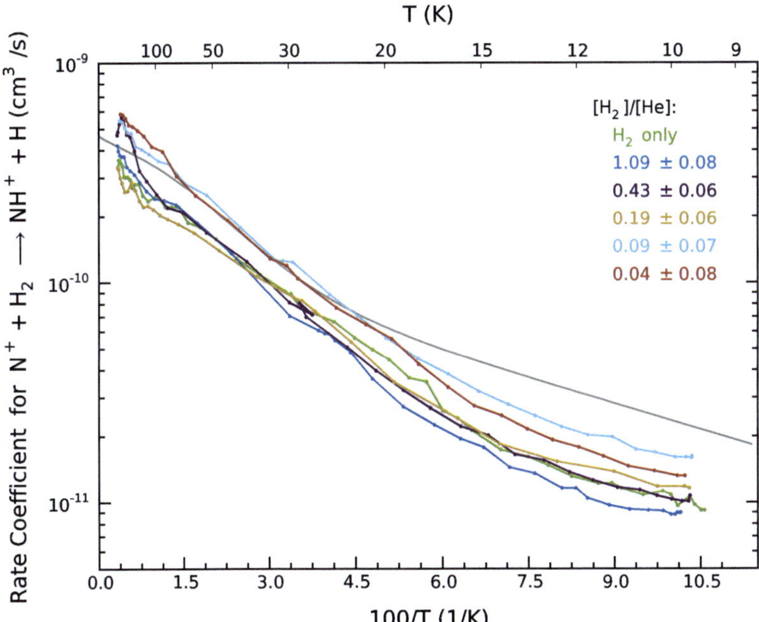

Figure 6.10: Rate coefficients for N^+ + n-H_2 $\rightarrow$ NH^+ + H for different amounts of Helium buffer gas in the trap. *Grey:* function (6.8) for n-H_2 (red line in Fig. 6.6). For the measurements shown here, He was admitted continuously to the 22-pole trap chamber as a mixture of n-H_2 and He. The ionization energy in the ion source was 25.5 eV and He was added to the source gas to improve the relaxation of the ions in the source.

The question remained, whether the amount of He in the trap influences the speed of the reaction or the deviation of the observed curves from a mono exponential reaction. A series of measurements was performed, to investigate this. As it is almost impossible to calculate the proper time-dependent number density of He in the trap, when it is injected by a pulsed valve, it was admitted to the trap continuously for these measurements.

Figure 6.10 shows the resulting temperature dependencies for different amounts of He in the trap. To obtain these measurements, the gas reservoir for the 22-pole trap chamber was first filled with n-H_2 only. Then for each next measurement, more He was added to this gas. To ensure good mixture of the two gases, at least twelve hours passed between the addition of more He and the next measurement series. Figure 6.10 reveals no trend of the curves for the possible influence of He. Figure 6.11 shows a comparison of N^+ time evolutions for the different measurement series at a nominal trap temperature of 223 K. More examples like this for other temperatures are given in the Appendix A.6.2. For better comparability of the measurements, the time axis of each curve was corrected for the change in the number density [n-H_2] (compare Equation (6.6)) and the N^+ axis for the change of the number of ions at $t = 0$, N_0. As could already be seen from Figure 6.10, no trend for the reaction speed is visible. Neither is an influence of the He density on the bending of the curves observable.

The measurements on the influence of the He density on reaction (6.8) re-

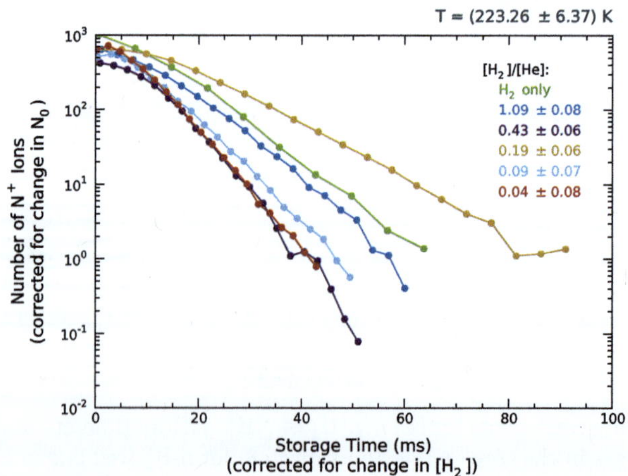

Figure 6.11: Decrease of the number of N^+ ions over the storage time in reaction with n-H_2 for different amounts of helium buffer gas in the trap at a nominal trap temperature of 223 K.

veal, that there still is an effect on the measurements, that is not understood so far. This causes problems with the reproducibility of the measurements. So far, the resulting rate coefficients should be considered to have an experimental uncertainty of $\approx 30\ \%$.

6.4.2 Effects of Different Ionization Energies on the Reaction System

At the beginning of the experiments on N^+, several different conditions for the production of ions were tested. In the storage ion source, N_2 is ionized by electron impact. All the ions produced in this process are then stored in the ion source until the next experimental cycle begins. During this storage time they can undergo reactive and relaxing collisions with the gases in the source. To improve the relaxation of the produced ions to the thermal equilibrium, helium can be added to the source.

The helium pressure in the source and the acceleration voltage of the electrons used for ionization have been varied to find the best conditions for the experiments. The original intention was to optimize the number of trapped N^+ ions. This number increases with the pressures of helium and nitrogen in the source and with the ionization energy. The effect of the He pressure is much stronger, than that of the N_2 pressure. The He collisions reduce the translational energy of the ions, making it more unlikely for them to leave the storage potentials.

The ionization energy was varied between 25.4 eV and 80 eV. For values below that range, the number of N^+ ions gets too small for acceptable pressures in the source. Table 6.3 gives the ionization energies of N and N_2, the binding energy of N_2, and the energies of the three lowest fine-structure states of N^+. Even the lowest ionization energy of 25.4 eV is sufficient, to excite the N^+ ions to the fine-structure states $^3P_1$ and $^3P_2$.

Table 6.3: Ionization energies of N and N_2, binding energy of N_2, and fine-structure states of N^+.

	N	N_2	N^+
Ionization Energy	14.5 eV[a]	15.6 eV[a]	
Binding Energy		9.8 eV[b]	
$^3P_0$			0 meV[c]
$^3P_1$			6.1 meV[c]
$^3P_2$			16.2 meV[c]

[a] from [54]   [b] from [55]   [c] from [50]

The pressures in the source have to be chosen under some restrictions. As explained above, a minimal flux through the leakage valves is required to reduce freezing effects during the measurement period. On the other hand, the pressures should not be too high for two reasons:

1. The pressure in the vacuum chamber has to stay low enough to avoid damage of the high voltage devices even at the time when the buffer gas is pulsed into the 22-pole trap.

2. The number of N^+ ions at $t = 0$ should not exceed 1000. For higher numbers of trapped ions, the Daly detector starts to loose sensitivity. For the observation of the decrease of N^+ ion with time, too high values of N_0 lead to a variation of the detection efficiency of the ions with the storage time. The result are curves that are too low for short times.

For these reasons, the first test measurements with different ionization energies were performed without He in the source. For each chosen ionization energy, the trapping and guiding potentials of the ion trap apparatus were optimized, to get the best possible storage of a feasible number of ions (preferably close to but not higher than 1000). Then a measurement series on the temperature dependence of the rate coefficient for the reaction of N^+ with either n-H_2 or p-H_2 was performed. The resulting $k_{f1}(T)$ curves in Figure 6.12 show large deviations from one another. Repeated measurements for n-H_2 showed good reproducibility. For p-H_2 several measurement series were taken at 25.4/25.5 eV. They also show rather large deviations from one another. The problem here might be, that the p-H_2 samples contained different amounts of o-H_2, due to slight changes in the production conditions for the fresh samples and due to back conversion for several days of measurements with one sample.

The most astonishing result from this measurements is the fact, that there is a maximum of the dependence of the rate coefficient on the ionization energy. For n-H_2 it was observed at 60 eV, for p-H_2 at 40 eV. For high temperatures, the $k_{f1}(T)$ curves reach their minimum at these ionization energies, for low temperatures they reach their maximum.

The most probable explanation for the observed correlation of the rate coefficient to the ionization energy is an effect of the fine-structure population of N^+. If the lowest level would react slower than the first excited level and the second excited level would not react at all, this could explain, why the rate coefficient is first increasing and then decreasing again towards higher ionization energies. The change in the slope of the curves in Figure 6.12 suggests, that not only the reaction rate for $\frac{1}{T} = 0$ depends on the fine-structure state, but also the activation energy. However, the question remains, why the fine-structure of N^+ should not be cooled to the thermal equilibrium by the buffer gas collisions in the trap.

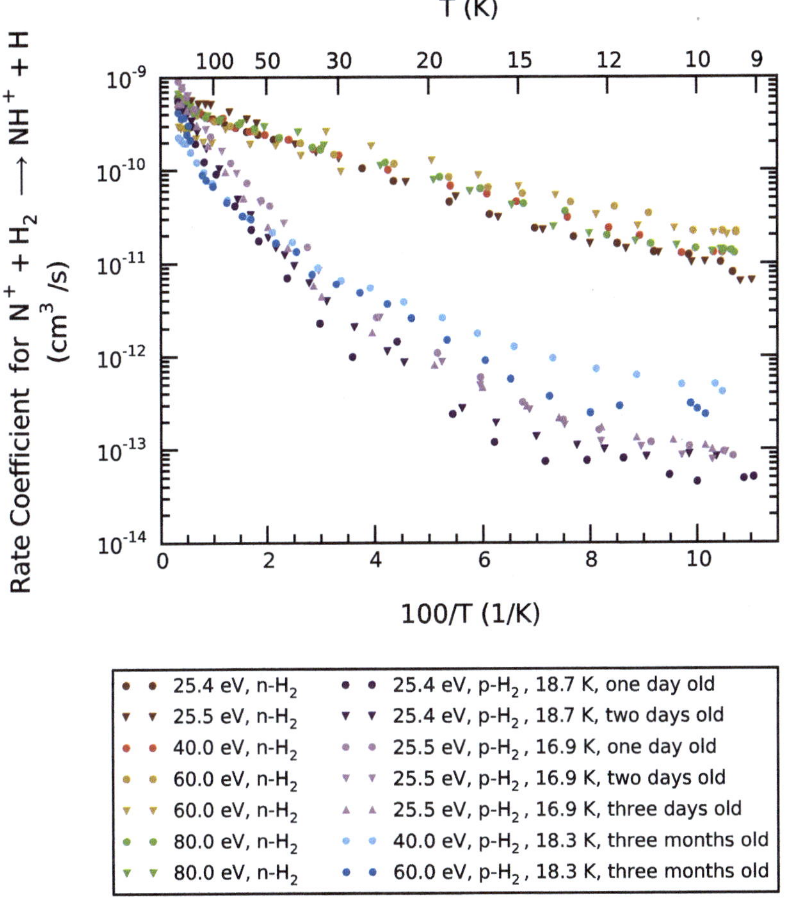

Figure 6.12: Temperature dependence of the rate coefficients for $N^+ + n\text{-}H_2 \rightarrow NH^+ + H$ for different ionization energies. T is the nominal trap temperature. The legend gives the ionization energies in eV and the properties of the investigated H_2 sample. For $p\text{-}H_2$, the maximal evaporation temperature and the time since the production of the samples are given as well.

At a later time, another series of measurements was performed on this effect. This time, He was added to the source gas and to the reactant n-H_2 gas in the trap (continuously instead of a He pulse). The trap was cooled down to ≈ 10 K and the ionization energy was changed without modifying any other experimental parameters intentionally. The results presented in Figure 6.13 show an experimental drift over the period of the measurement series. When the rate coefficients are corrected for this drift (for this purpose, two measurements were performed at 25.5 eV), no trend remains.

This strengthens the theory, that the effects of the ionization energy observed before are due to the influence of the ionization energy on the fine-structure population of N^+. The constant presence of He in the trap and in the source increases the relaxation processes, thereby reducing the influence of the ionization energy on the rate coefficient. To verify this, it is recommended to repeat the measurements with reduced amounts of He in the source and the trap, to check whether the ionization energy influence increases.

Another effect was observed, when the ionization energy was varied. For values above 40 eV, the time evolution of the N^+ ions showed a constant background on mass 14. Figure 6.14 shows the fit results for the background counts from the measurement series whose rate coefficients are presented in Figure 6.13. The background counts seem to increase linearly with the ionization energy in the range above 40 eV. There are two possible explanations for this.

When the ionization energy reaches a certain level, the ions in the source might gain enough translational energy to overcome the potentials that usually should hinder them from passing the 22-pole trap during the measurement cycle.

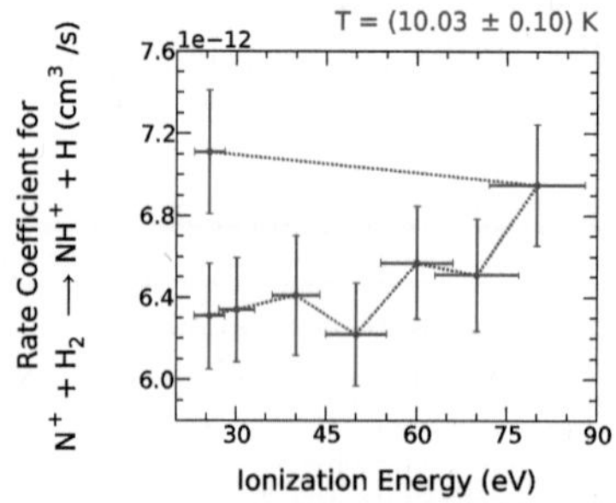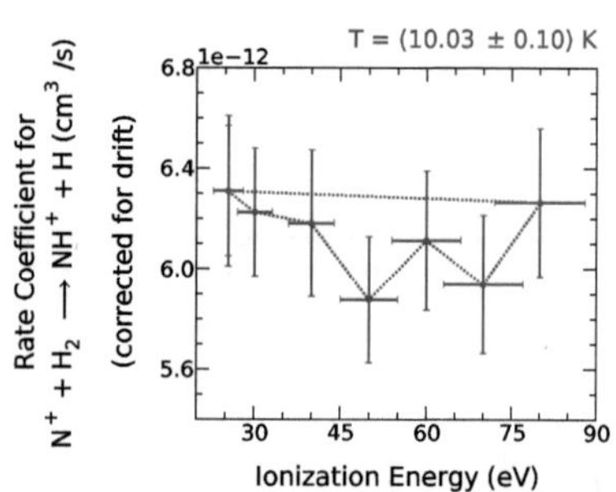

Figure 6.13: Rate coefficients for $N^+ + $ n-$H_2 \to NH^+ + H$ for different ionization energies at a nominal trap temperature of 10 K.
Left: The dotted lines connect the values in the order they were measured (from low ionization energies to high ones and back to the lowest value).
Right: The same results as on the left, corrected for the experimental drift.

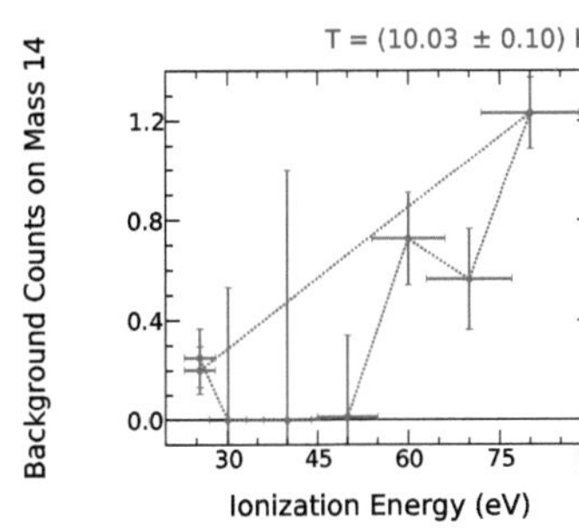

Figure 6.14: Background counts on mass 14 for different ionization energies at a nominal trap temperature of 10 K. The dotted lines connect the values in the order, they were measured (from low ionization energies to high ones and back to the lowest value). The experimental drift between the first and the last measurement is negligible.

Then a constant rate of ions should hit the detector at all times. To verify this theory, the opening time of the gate (i.e. the time period in that ions are counted after opening the exit of the trap) was varied. The background counts increased with longer gate times, but not proportional. Thus, only part of the background is probably due to the explained effect.

Another theory was, that some molecules or atoms might travel from the source to the trap in highly exited metastable states. They would then be ionized in the 22-pole trap by collisions with ions. To test this theory, the mass selection of the first quadrupole was set to mass 15 instead of mass 14. The background on mass 14 vanished completely. Thus, this explanation cannot be true.

So far, there are no other theories concerning the origin of the background.

6.4.3 Influence of the Amplitude of the Trapping RF

The amplitude of the radio frequency voltage applied to the 22-pole has to be chosen carefully to avoid wrong results for the forward rate coefficient of reaction (6.1). Two effects have to be considered here:

1. For too low amplitudes of the RF, the N^+ ions are not trapped properly. The resulting loss of ions with the storage time seemingly increases the rate coefficient, which is derived from the time constant with that the number of N^+ ions in the trap decreases.

2. For too high amplitudes, the ions gain energy from the electrical field. This energy makes it more likely to overcome the endothermicity of the reaction, resulting in too high values for the rate coefficient as well.

The best conditions to observe these effects are at low temperatures and with only p-H_2 present in the trap. Under such conditions, the above mentioned effects will be strongest.

During the measurements, the amplitude of the RF cannot be measured directly. Only the dialed value on the frequency generator RF_{dial} is known. From this, the actual amplitude RF_{amp} has to be derived by calibration measurements, resulting in:

$$RF_{\text{amp}} = (205.45 \pm 3.01)\ \text{V} \cdot RF_{\text{dial}} - (471.15 \pm 7.29)\ \text{V}, \qquad (6.15)$$

for $RF_{\text{dial}} > 2.4$. The calibration measurements are shown in the Appendix A.6.3.

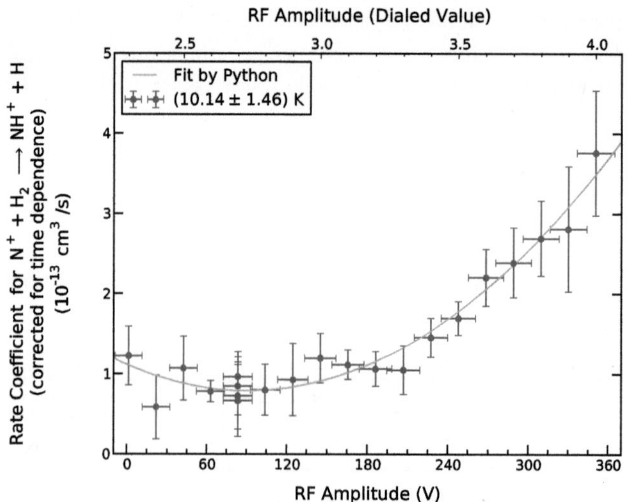

Figure 6.15: Influence of the RF amplitude on the observed rate coefficient. For low amplitudes, the N^+ ions can leave the trap and the rate coefficient seems to increase. For high amplitudes, the electric fields heat the ions and thereby increase the reaction rate. The measurements were performed with p-H_2 and at ≈ 10 K. Under these conditions the effect of the RF is strongest. Fit function (6.16) is shown as *solid line*. The measurements presented here were corrected for experimental drifts (compare Figure 6.13).

Figure 6.15 shows the results for the dependence of the observed rate coefficient on the RF amplitude. It can be fitted by the function

$$k_{f1}\left(RF_{\mathrm{amp}}\right) \;=\; a\left(RF_{\mathrm{amp}} - b\right)^2 + c\,, \tag{6.16}$$

$$a \;=\; (3.995 \pm 0.433) \times 10^{-18}\ \frac{\mathrm{cm}^3}{\mathrm{Vs}}\,,$$

$$b \;=\; (90.53 \pm 10.63)\ \mathrm{V}\,,$$

$$c \;=\; (7.923 \pm 0.518) \times 10^{-14}\ \frac{\mathrm{cm}^3}{\mathrm{s}}\,.$$

The measurements presented in Fig. 6.15 showed a drift with the time since the p-H_2 valve was opened. The observed rate coefficients k_{f1} were corrected for this drift. For this purpose several measurements were performed at a dialed RF amplitude value at the RF generator of 2.7, where the minimum of the curve (6.16) was expected from earlier tests. The drift was then fitted by a linear function and the late time values for k_{f1} were corrected to the early time results. The minimum of the fitted function was found to be at:

$$RF_{\mathrm{amp}} \;=\; (90.53 \pm 10.63)\ \mathrm{V} \tag{6.17}$$
$$RF_{\mathrm{dial}} \;=\; 2.7339 \pm 0.0744 \tag{6.18}$$

For further measurements, the RF was therefore set to 2.7.

6.4.4 Conclusion on the Influence of Experimental Parameters.

The reaction system $N^+ + H_2$ is very sensitive to any form of kinetic energy of the reaction partners as well as to the HD-to-H_2 ratio of the reactant gas. Hence, many experimental parameters were found to influence the observed reaction rate, which usually are far less crucial in other ion trap experiments.

The reproducibility of the experimental results could be improved by choosing the same cycle time for all measurements, not changing the electric guiding and trapping potentials of the ion trap experiment and setting the current of the ionization pressure gauge to 1 mA. The influence of the trapping RF is minimized by operating the RF generator at an amplitude of ≈ 90 V (i.e. 2.7 dialed at the potentiometer) and optimizing the RF frequency for each temperature.

The influence of the buffer gas density in the ion trap and the number densities of the precursor and the buffer gas in the ion source need to be reinvestigated. All three number densities are expected to crucially influence the fine-structure population of the N^+ ions. For future measurements, it is necessary to improve the precision with which these values can be set. As a first step, a more precise pressure gauge[1] will be installed at the gas inlet system of the ion trap. This will allow to determine the partial pressures of H_2 and He in mixtures of both gases with much smaller uncertainties than possible so far. As another step, it is recommended to replace the leakage valves for the gas inlets to ion source and ion trap by more reliable ones (new leakage valves are usually capable of the required fluxes, the problem discussed here is most likely due to the age of the current valves).

6.5 Results for the Purity of p-H_2

Different test methods have been applied to investigate the ortho-to-para ratio of gas samples produced in the para hydrogen converter.

The reaction equilibrium of $H_2D^+ + H_2 \leftrightharpoons H_3^+ + HD$ at 10 K is very sensitive on the ortho fraction of the reactant hydrogen, but also on its HD amount. In both production methods of p-H_2, the HD amount is decreased compared to the natural amount. This effect is strongest for the continuous flow method. Freezing effects of HD have also been observed in the ion trap. Additionally, a proper mathematical description of the time dependence of the $\frac{H_2D^+}{H_3^+}$ ratio for the back conversion of para hydrogen to normal hydrogen in a storage bottle is lacking. For both production methods, estimates for the purity of p-H_2 have been made on the best guess for the values of all involved experimental parameters. They are summarized together with the results from Raman spectroscopy and the reaction $N^+ + H_2 \leftrightharpoons NH^+ + H$ in Table 6.4. These results include the estimated $\frac{[HD]}{[H_2]}$ ratio. An overestimation of this ratio would result in an overestimation of the ortho fraction (see Equation (5.21)).

Raman spectroscopy allows the direct comparison of level population for ortho and para states. This was expected to be very precise for ortho fractions of some percent, but it comes with a lower detection limit. This detection limit was found to be above the ortho fractions we expect for the usual production temperatures of para hydrogen. Attempts were made to investigate the time constant for the back conversion in the glass cell. They were hindered by experimental drifts of the Raman spectrometer who's origin is not yet determined. Thus, previous tests [20] on the back conversion rate in the Teflon coated storage bottle were not yet repeated.

The reaction system $N^+ + H_2 \leftrightharpoons NH^+ + H$ was investigated in great detail. It turned out to be very sensitive on various experimental parameters, causing trouble with the reproducibility of the results. The deuterium contents of the reactant gas influence this reaction system as well, as HD reacts with N^+ with a rate that is close to the one for the reaction with o-H_2. Thus, an underestimation

---

[1]Swagelok, PTU Series Ultrahigh-Purity Pressure Transducer, instead of
   Swagelok, PGU-50-PC100-L-4FSF Manometer

Table 6.4: Results for the ortho fraction of different para hydrogen samples.

Test Method	Production Method	Production Temperature	$f = \frac{[o\text{-}H_2]}{[H_2]} \times 10^4$	
			Expected	Measured
$\frac{H_2D^+}{H_3^+}$	Continuous Flow	18.8 K	9.23	$f < 14.8$
	Freeze Out	20.8 K	22.30	$2.78 < f < 15.8$
Raman	Freeze Out	20.0 K	16.01	$f < 50$
$N^+ + H_2$	Freeze Out	17.3 K	4.17	$f < (15.0 \pm 7.5)$

of the HD amount would result in an overestimation of the ortho fraction. Measurements on the back conversion rate of p-H$_2$ in the Teflon bottle were planned. The intention was to perform experiments in the ion trap and in the Raman spectrometer simultaneously. As both experiments showed drifts that are not (fully) understood so far, measurements on the back conversion in the Teflon bottle (which would take three months) were not yet performed.

Both reaction systems $H_2D^+ + H_2 \leftrightharpoons H_3^+ + HD$ and $N^+ + H_2 \leftrightharpoons NH^+ + H$ are influenced not only by the ortho-to-para ratio of the reactant gas, but also by its HD fraction. Luckily the influence of HD is opposite for the two systems. For the H_3^+ system HD kind of behaves like p-H$_2$, for the N^+ system like o-H$_2$. For future tests, the production temperatures of the p-H$_2$ sample should be chosen equal for both test reactions. This will hopefully enable us to derive more precise values for the ortho fraction and maybe even for the HD fraction of the p-H$_2$ samples.

Chapter 7

Spectroscopy of CH_2D^+

The ion CH_3^+ and its isotopologues are of astrophysical interest, because they transfer deuterium from HD into larger molecules, starting with the exothermic reactions:

$$CH_3^+ \; + HD \; \rightleftharpoons \; CH_2D^+ + H_2 \tag{7.1}$$

$$CH_2D^+ + HD \; \rightleftharpoons \; CD_2H^+ + H_2 \tag{7.2}$$

$$CD_2H^+ + HD \; \rightleftharpoons \; CD_3^+ \; + H_2 \tag{7.3}$$

As ions are very reactive and it is energetically preferable to have deuterium in the larger one of two reaction partners, deuterium is easily transferred to heavier molecules from the CH_3^+ isotopologues.

For this reason, the CH_3^+ isotopologues have been investigated spectroscopically in the laboratories of Takeshi Oka [56], [57] and there were continued efforts to detect them in the interstellar medium.

At the beginning of this work, the group of Evelyne Roueff observed lines in Orion IRc2 [58] that were good candidates for CH_2D^+. The deviations of the observed line positions in space from the predictions based on infrared measurements by Rösslein et al 1991 [56] and Jagod et al 1992a [57] however, were too large for a confirmation of these detections. This resulted in new efforts of several groups to improve the precision of predictions for pure rotational transitions of CH_2D^+. Parallel to the first publication of the results presented in the following [14], Takayoshi Amano reported the detection of four pure rotational transitions of CH_2D^+ [33]. A more detailed description of the measurements presented here has been published in [13].

7.1   The CH_2D^+ Ion

CH_2D^+ is a planar asymmetric rotor with C_{2v} symmetry. The orientation of the principle axes in this molecule is shown in figure 7.1. The symmetry axis of

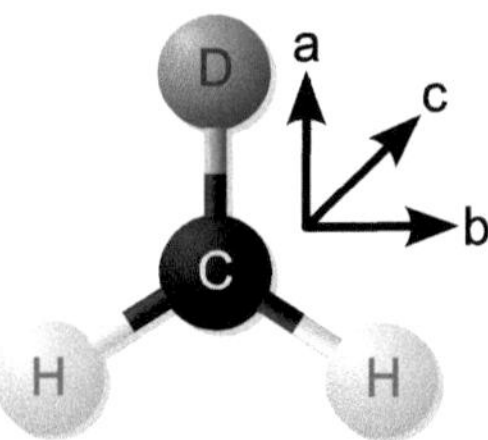

Figure 7.1: Structure and principle axes of the CH_2D^+ ion. It is a planar asymmetric rotor with C_{2v} symmetry.

CH_2D^+ is the a-axis, the dipole moment μ_a is 0.329 debye [57].

In this work two vibrational bands of CH_2D^+ have been investigated:

1. The ν_1 vibration is the totally symmetric (A_1) stretching vibration. The vibrational dipole moment is oriented along the a-axis leading to a-type rotational selection rules for the transitions:
$\Delta K_a = 0, \pm 2, \ldots$ and $\Delta K_c = \pm 1, \pm 3, \ldots$ (K_i is the projection of the total angular momentum $\vec{J}$ on the axis i, see Figures 7.1 and 7.2).

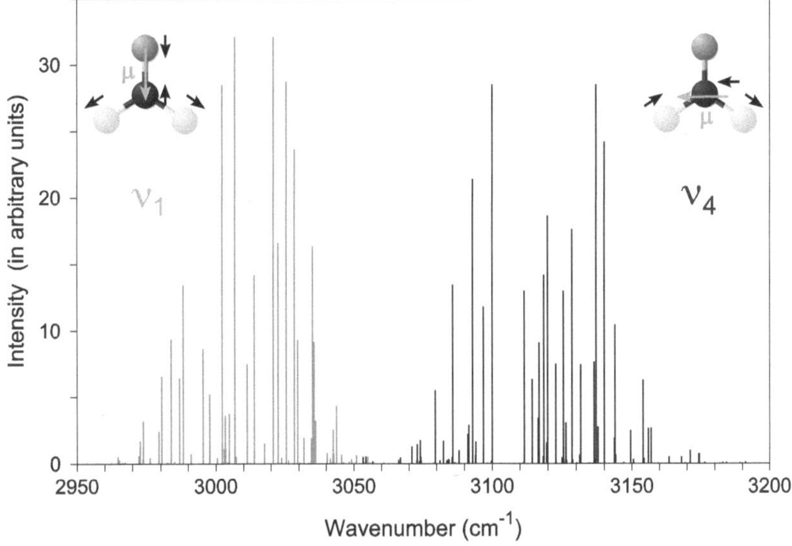

Figure 7.2: Simulation of the ν_1 and ν_4 vibrational bands of CH_2D^+ at 30 K from pgopher [59]. Also shown are schematics of the vibrational motions of CH_2D^+ and the vibrational dipole moments μ in both modes.

2. The ν_4 vibration is an antisymmetric (B_1) stretching vibration. The vibrational dipole moment is oriented along the b-axis leading to b-type rotational selection rules for the transitions:
   $\Delta K_a = \pm 1, \pm 3, \ldots$ and $\Delta K_c = \pm 1, \pm 3, \ldots$

More details on the spectroscopy of planar asymmetric rotors can be found in [48]. Figure 7.2 shows a `pgopher` [59] simulation at 30 K of both vibrational bands and schematics of the vibrational motions excited in either state. The transitions of ν_1 and ν_4 lie in the range of 2900 to 3200 cm^{-1}.

Figure 7.3 shows examples of the selection rules for both vibrational transitions. Since the ion contains two identical hydrogen nuclei, nuclear spin statistics have to be considered. Like the hydrogen molecule (see also Chapter 5), CH_2D^+ can exist in an ortho and a para state. I.e. the two nuclear spins of $\frac{1}{2}$ of the individual H atoms can couple either in a symmetric or an anti symmetric way. As there are three symmetric linear combinations of the spins and only one antisym-

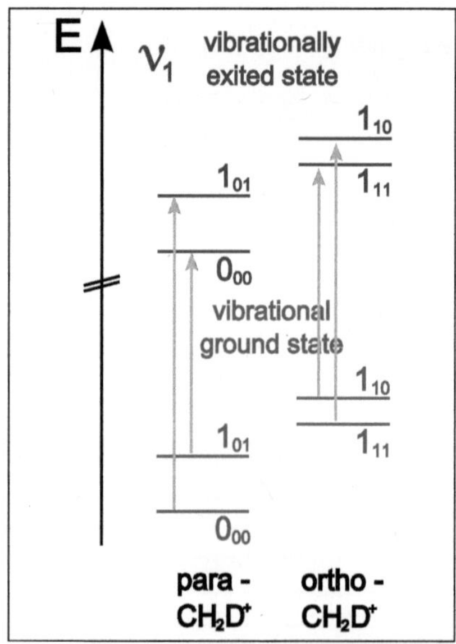

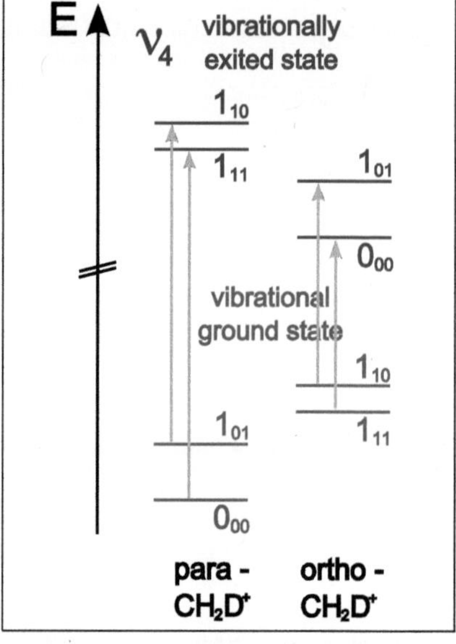

Figure 7.3: Selection rules for transitions in the ν_1 and ν_4 bands of CH_2D^+. For ν_1, a-type transitions are allowed, leading to the selection rules $\Delta K_a = 0, \pm 2, \ldots$ and $\Delta K_c = \pm 1, \pm 3, \ldots$. The ν_4 branch is formed by b-type transitions with the selection rules $\Delta K_a = \pm 1, \pm 3, \ldots$ and $\Delta K_c = \pm 1, \pm 3, \ldots$. Only transitions with $J \leq 1$ are shown here.

metric one, the ortho to para ratio is 3:1. As for H_2, the nuclear spin configuration of CH_2D^+ determines, which rovibrational levels can be occupied.

7.2 LIR Spectroscopy of CH_2D^+

The spectroscopic investigations of CH_2D^+ were performed in the LIRTrap apparatus. As radiation source an optical parametric oscillator (homebuilt by Jürgen Krieg, see Section 7.2.2 and [60]) was combined with the ion trap.

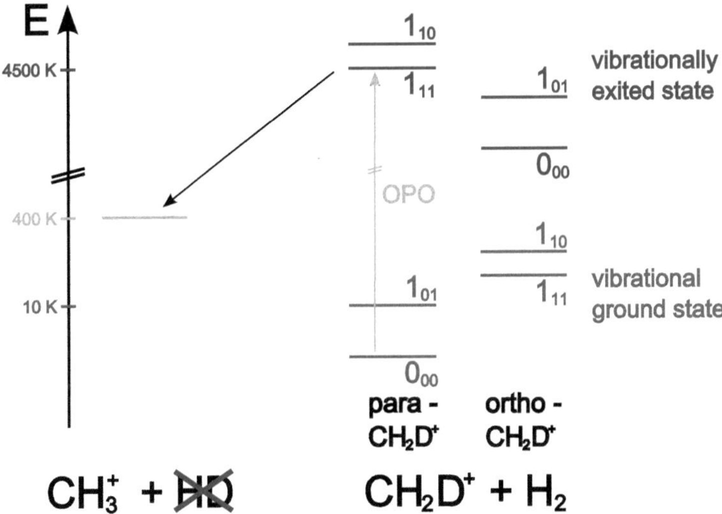

Figure 7.4: Schematic of the LIR principle for CH_2D^+, from [13]: Ions in the vibrational ground state *(blue)* do not have enough energy to react to CH_3^+. When an ion is laser pumped into a vibrationally excited state *(red)*, it can react with H_2 and form CH_3^+. If HD molecules were present in the trap, the product ions CH_3^+ would react back to CH_2D^+. This reaction can be neglected, as only the very small natural abundance of HD in H_2 is admitted to the trap.
As CH_2D^+ contains two H atoms, it comes in two nuclear spin configurations. Radiational transitions between the ortho and the para state are forbidden by conservation of rotational momentum (nuclear spin, in this case). Radiational transitions within the ortho or para ladder are allowed as long as they fulfill the selection rules.

7.2.1   Production and Trapping of CH_2D^+

Singly deuterated methane[1] (CH_3D) was ionized for the production of CH_2D^+. In order to cool the produced ions, helium gas was injected to the ion source as well. This enhances the thermalisation of the produced ions to room temperature after the electron impact ionization. To prevent ionization of He, the ionzation energy was kept below 24 eV (ionization energy $\approx$ 24.6 eV).

The produced ions were mass filtered for CH_2D^+ (mass 16 u) before entering the trap. In the cold 22-pole trap, the ions were cooled down by an intense helium pulse and then stored together with the reaction gas H_2 for 480 ms. They were subject to laser radiation for the whole storage time. When a CH_2D^+ ion absorbs an infrared photon, it has enough energy to react with H_2 to CH_3^+ as in the backward direction of (7.1). This LIR process is depicted in Figure 7.4 and is explained in more detail in Chapter 4.4.

After the storage time, the ion cloud was extracted from the trap. Before hitting the Daly detector the ions were mass selected for CH_3^+ (mass 15 u). A more thorough description of the trapping process can be found in Chapter 4.1.

7.2.2   Optical Parametric Oscillator

The radiation source for the spectroscopic investigations of CH_2D^+ is an optical parametric oscillator (OPO) operating in the wavelength region around 3 μm. It was built by Jürgen Krieg and is described thoroughly in his PhD thesis [60]. A similar setup for wavelengths around 5 μm is described in [61].

Figure 7.5 shows a schematic of the OPO system. In the lithium niobate crystal[2], power from a pump laser[3] is transferred to two other beams, called signal and idler by effects of nonlinear optics. The wavelengths of those beams are widely tuneable by the operation parameters of the OPO (coarse tuning: crystal temperature[4], poling periode of the crystal[5]; fine tuning: continuous tuning of the pump laser[6]).

The signal beam (1.45 to 1.85 μm) is kept resonant in a high-finesse ring cavity to enhance the conversion process. The idler beam (2.5 to 4.0 μm, i.e. 2500 to

---

[1]Cambrigde Isotope Laboratories, Inc.: Methane (D1, 98 %)

[2]5 % MgO-doped periodically poled $LiNbO_3$

[3]Innolight MOP series, Nd:YAG 10 W optical power at 1064 nm

[4]The crystal is held in an oven at temperatures between 50 and 200 °C.

[5]Seven distinct poling periods between 28.5 and 31.5 μm are available in the crystal for phase matching of the OPO process.

[6]15 GHz mode-hop-free, 80 GHz total

[7]Optical powers of more than 800 mW have been observed.

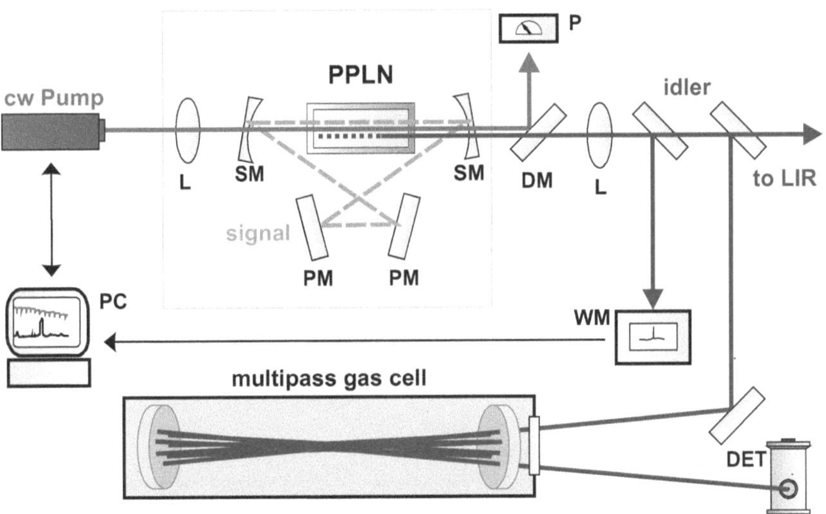

Figure 7.5: Schematic of the homebuilt OPO system, from [13].
Abbreviations: L, lens; SM, spherical mirror; PPLN, periodically poled lithium niobate; PM, plane mirror; WM, wavemeter; DM, dichroic mirror; cw, continuous wave; and P, power meter. For details, see text.

4000 cm^{-1}) is used for the experiments. The optical power of the idler[7] was kept around 100 mW by limiting the pump power for the experiments.

7.2.3 Frequency Determination

The frequency of the OPO was varied slowly while the LIR cycle was repeated continuously. That way a spectrum of CH_2D^+ was recorded as frequency dependent increase of the number of CH_3^+ ions. For frequency calibration part of the radiation was guided through a Herriott-type absorption cell[8], filled with either carbonyl sulfide (OCS) or air (for H_2O lines) and another part into a wavemeter[9].

In order to keep the frequency uncertainties resulting from the frequency calibration comparable to or smaller than the uncertainties of the fit of the measured lines, only calibration lines with an accuracy better than 1.3×10^{-4} cm^{-1} were used. They were connected to the measured CH_2D^+ lines in mode hop free scans (see Figure 7.6), which limited the number of CH_2D^+ lines that could be measured with sufficient quality of the frequency calibration to 112.

---

[8]total absorption path length: 50 m
[9]Bristol Instruments, model 621-A IR

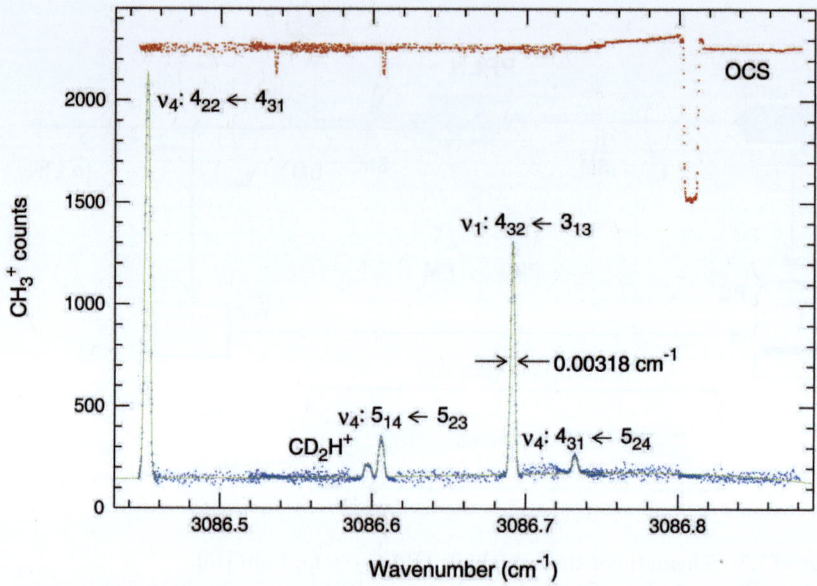

Figure 7.6: Example of a CH_2D^+ spectrum observed with LIR spectroscopy, from [13].
Blue: When the OPO is tuned on a rovibrational transition of CH_2D^+ the number of CH_3^+ ions increases. There is also one line visible, that belongs to CD_2H^+. Its origin is explained in the text.
Green: Saturated Gaussian lines were fitted manually to the data.
Red: Parallel to the ion trap measurements an OCS reference spectrum was taken for frequency calibration (given in arbitrary intensity units).

It was necessary to use molecular calibration lines additional to the wavemeter for frequency determination, because the wavemeter results have an offset (up to ≈ 0.003 cm^{-1}), that is - among others - depending on the beam alignment and therefore on the OPO mode, as mode hops always result in slight changes in the beam path. Those changes are small enough to not hamper the LIR measurements, but cannot be neglected in the frequency determination. Also very small drifts of the wavemeter measurements over scans of wide frequency ranges were observed, when the device was used together with other lasers. These drifts however do not affect the results presented here, as the frequency ranges in each scan are very small (< 0.5 cm^{-1}).

For calibration, two gases were used: OCS and H_2O. For the strong lines of OCS, calculated line positions from NIST[10] [54], based on measurements by

Saupe [62], were used. For some weaker OCS lines the positions were taken from Guelachvili and Rao [63], after correcting for an offset which the lines in this reference show, compared to the more accurate NIST lines. OCS could be used as reference gas only in the range 3066 to 3120 cm^{-1}, because of the required accuracies of the line positions. By using water lines, the measurement range could be expanded to 2954 to 3133 cm^{-1}. The line positions for H_2O were taken from Toth [64]. For the weaker H_2O lines, the Herriott cell was filled with laboratory air at low pressure, while for the stronger lines, the amount of water in the laboratory air was sufficient without the use of a multi reflection cell.

7.2.4 Spectra of CH_2D^+

Figure 7.6 shows an example of a CH_2D^+ spectrum, together with the simultaneously recorded OCS reference spectrum. Usually the mass selection before and after the trapping process guaranties for spectra of only one ion type. The spectra of CH_2D^+ recorded in this work show also three lines of CD_2H^+. Those lines are probably formed via the admission of CD_2^+ ions into the trap, that cannot be distinguished from CH_2D^+ in the quadrupole mass filter. In the trap CD_2^+ forms CD_2H^+ in collisions with H_2 via reaction (7.4). The CD_2H^+ ions can then react with H_2 to form CH_3^+ via reaction (7.5). Since this reaction is endothermic, it happens only after the ion absorbed a photon. As there are only very few CD_2H^+ ions formed in the trap, only three lines with a large Einstein B-coefficient appeared in the spectra. Those lines can easily be distinguished from the CH_2D^+ lines due to their strong deviation from the Gaussian line profile.

$$CD_2^+ \quad + H_2 \qquad \rightarrow \quad CD_2H^+ + H \qquad (7.4)$$
$$CD_2H^+ + H_2 + h\nu \quad \rightarrow \quad CH_3^+ \quad + D_2 \qquad (7.5)$$

Figure 7.7 shows an enlarged view of the two lines observed near 3086.6 cm^{-1}. The smaller line on the left is one of the three observed CD_2H^+ lines. Both lines show deviations from the Gaussian line profile, but the deviations are much stronger for the CD_2H^+ line. The possible origin of this deviations is discussed in more detail in Chapter 8.

7.3 Spectroscopic Results

In this work, 112 rovibrational transitions of CH_2D^+ have been observed. 111 of those lines were included in a fit with `spfit` [65] (one line had too low quality to be included in the fit). The results are given in table 7.1 and were published in [14] and [13].

---

[10]National Institute of Standards and Technology

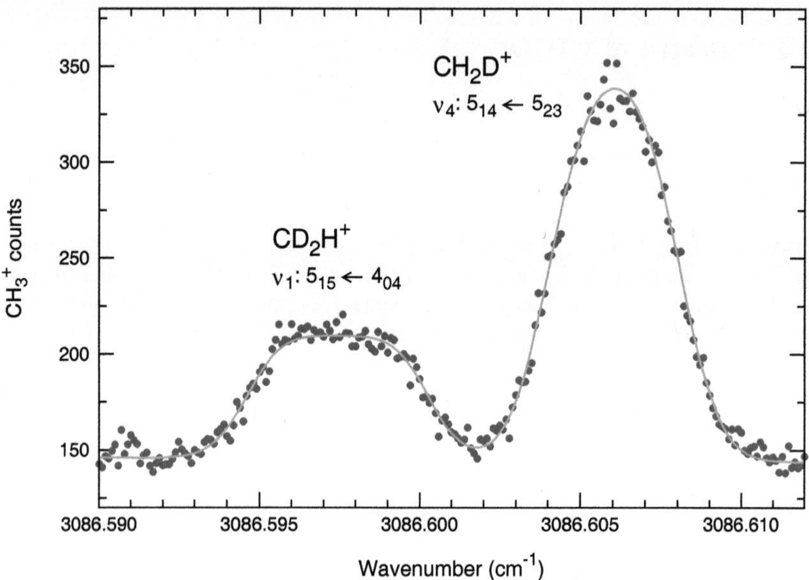

Figure 7.7: Comparison of a CD_2H^+ (left) and a CH_2D^+ (right) line (enlarged view from figure 7.6), from [13]. The line of CD_2H^+ is much flatter on top than the CH_2D^+ line. This saturation effect will be discussed in detail in Chapter 8.

Table 7.1: Best fit molecular parameters in ground and excited vibrational states of CH_2D^+ based only on the lines measured in this work, given in cm^{-1}. In total 35 parameters have been used to fit 111 transitions with `spfit` [65] using a standard Watson-type Hamiltonian in A-reduction. The numbers in parentheses give the standard deviation of the last digits.

Parameter		Ground State	ν_4	ν_1
ν			3105.840584(52)	3004.764593(75)
A		9.368512(25)	9.270327(33)	9.210940(47)
B		5.771331(21)	5.732475(15)	5.757158(32)
C		3.525266(21)	3.502936(22)	3.494216(31)
Δ_J	$\times 10^3$	0.12052(31)	0.12249(43)	0.11829(36)
Δ_{JK}	$\times 10^3$	0.3603(17)	0.3450(38)	0.3586(36)
Δ_K	$\times 10^3$	0.2601(24)	0.2988(20)	0.2690(57)
δ_J	$\times 10^3$	0.04772(90)	0.04596(29)	0.0490(12)
δ_K	$\times 10^3$	0.4193(37)	0.3952(48)	0.4105(48)
Φ_J	$\times 10^6$			
Φ_{JK}	$\times 10^6$		0.209(95)	-0.171(89)
Φ_{KJ}	$\times 10^6$	-0.35(12)		0.91(26)
Φ_K	$\times 10^6$			
ϕ_J	$\times 10^6$	0.024(21)		0.045(20)
ϕ_{JK}	$\times 10^6$	0.61(13)	-0.26 (18)	0.46 (11)
ϕ_K	$\times 10^6$			

Chapter 8

Line Profiles Observed with LIRTrap

The spectral lines measured by LIR spectroscopy often show significant deviations from the Gaussian line profile (see e.g. Chapter 7). The origin of this deviation, the proper mathematical description of the observed profiles (see also [13]) and first steps towards the quantitative understanding of the involved processes will be described in the following.

8.1   Theory

In spectroscopic ion trap experiments the ions are stored in the trap with the neutral reaction gas for several hundred milliseconds. For most or all of the time they are subject to laser radiation. The reaction system evolves towards its equilibrium, which is influenced by the frequency of the radiation. Hence, the spectra taken with LIR show the frequency dependence of the status of the observed endothermic reaction at a specific reaction time. They do not show a momentary picture of the velocity distribution of the ions, although the velocity distribution of course influences the frequency dependence. In the following we assume the irradiation time to be the same as the reaction (i.e. storage) time, meaning that the laser radiation is present in the trap at all times of the experiment.

We consider an example reaction system $X^+ + U \rightleftharpoons Y^+ + V$ which is endothermic in the forward direction. The number of product ions N_Y will depend on the laser frequency ν and the reaction time t as

$$N_Y(t, \nu) = N_0 \left(1 - e^{-K(\nu)t}\right) , \qquad (8.1)$$

where N_0 is the number of parent ions X^+ available for the reaction and $K(\nu)$ is an effective reaction rate, depending on several parameters described in the following.

During the storage time the ions undergo collisions with the neutral reaction gas and with the buffer gas (helium). Inelastic collisions keep up a Maxwell-Boltzmann velocity distribution at any time of the experiment. This leads to a Gaussian term describing the frequency dependence of the photon absorption probability, and the reaction rate becomes

$$K(\nu) = A e^{-\frac{mc^2}{2kT}\left(\frac{\nu - \nu_0}{\nu_0}\right)^2} . \tag{8.2}$$

m is the mass and T the temperature of the parent ions X^+, c is the speed of light and k the Boltzmann constant. The parameter A includes all effects that do not depend on the frequency or the reaction time. For weak radiation fields it will be proportional to the Einstein B coefficient of the observed transition and to the number of photons available for absorption and therefore to the laser power. A will increase with the reactivity of the excited ions and with the repopulation rate of the lower state (from the other internal states of X^+). It will be reduced by any competitive reactions in the real experiment and by quenching, i.e. inelastic collisions leading to relaxation to lower internal levels.

We have to add a background term $C(\nu)$ to equation (8.1), since the ions enter the trap hot and usually have enough energy to react and form the products before they are cooled down in collisions with the buffer gas. The frequency dependence of C corrects for slight systematic drifts during the experiments.

In total the time and frequency dependence of the number of product ions can be described as

$$N_Y(t, \nu) = N_0 \left(1 - e^{-Ate^{-\frac{(\nu - \nu_0)^2}{2\sigma^2}}} \right) + C(\nu) , \tag{8.3}$$

where the standard deviation σ is $\nu_0 \sqrt{\frac{kT}{mc^2}}$. For very small values of $A \cdot t$ this can be expanded into a Taylor series, leading to a Gaussian line profile

$$N_Y(t, \nu) = N_0 \left(Ate^{-\frac{(\nu - \nu_0)^2}{2\sigma^2}} + \dots \right) + C(\nu) . \tag{8.4}$$

Therefore, weak absorption and short irradiation times lead to LIR lines with Gaussian profiles.

8.2 Numerical Simulations with Python

Equation (8.3) has the form of Lambert-Beer's law combined with a Gaussian term. This function is well known to describe the line profiles in absorption experiments where the number of photons is limited compared to the number of absorbing molecules. The absorption of a photon can be viewed as a collisional process. In this picture there is no difference whether the number of photons or the number of absorbing particles is limited. Therefore, functions of the form (8.3) fit lines saturated by either process. The quantitative description of the processes for a limited number of photons is well understood, but for a limited number of absorbing particles, the description is only qualitative so far. The parameters N_0 and A are varied to achieve the best agreement between measurement and calculated line profile. However, their dependence on experimental conditions and on properties of the investigated reaction system is unknown. In order to improve the understanding of the LIR processes, numerical simulations were performed based on the reaction system

$$\mathrm{H_3^+ + HD \rightleftharpoons H_2D^+ + H_2 + 232\ K}\ . \tag{8.5}$$

This reaction system was chosen as basis of the model, because it is the best possible test case for laboratory investigations of the saturation effects. Spectroscopy of $\mathrm{H_2D^+}$ has been performed in the LIRTrap before with high quality results [18, 34]. The spectra of this ion are well understood and a suitable IR laser[1] together with an amplifier[2] is available. Earlier measurements [7] showed, that it is possible to obtain saturated as well as Gaussian line profiles for $\mathrm{H_2D^+}$ with this laser.

8.2.1   Simplified Model for Numerical Simulations

The simplest model to numerically simulate the LIR process contains three energy levels and three rates for transitions between those levels. Figure 8.1 shows a schematic of that reaction system. The following effects are included: laser pumping, reaction of excited ions with the neutral reactant gas, and frequency dependence of the number of reacting ions.

Laser Pumping:

The process of laser induced reactions starts with an ion absorbing a photon, so we need to consider a lower state $|\mathrm{L}\rangle$, representing a rotational level in the vibrational ground state, and an upper state $|\mathrm{U}\rangle$, representing a rotational level in a vibrationally excited state (see Figure 8.1). Ions can change from one state to the other via laser pumping. The rates K_{Pu} and K_{Pd} for pumping the ions up and down are given by

---

[1]Agilent Technologies, 81600B Tunable Laser Source Family Model #200, EC-Laser InGaAsP, wavelength range 1440 to 1640 nm, max output power <15 mW

[2]Calmar Optcom, Model: AMP-PM-15

$$K_{Pd} \;=\; \frac{B_{emis} \cdot p_{Laser}}{\sigma \cdot a \cdot c} \;, \tag{8.6}$$

$$=\; \frac{A_{emis} \cdot p_{Laser} \lambda^4}{8\pi \cdot h \cdot a \cdot c} \sqrt{\frac{m}{k_B T}} \;, \tag{8.7}$$

$$K_{Pu} \;=\; \frac{B_{abs} \cdot p_{Laser}}{\sigma \cdot a \cdot c} \;, \tag{8.8}$$

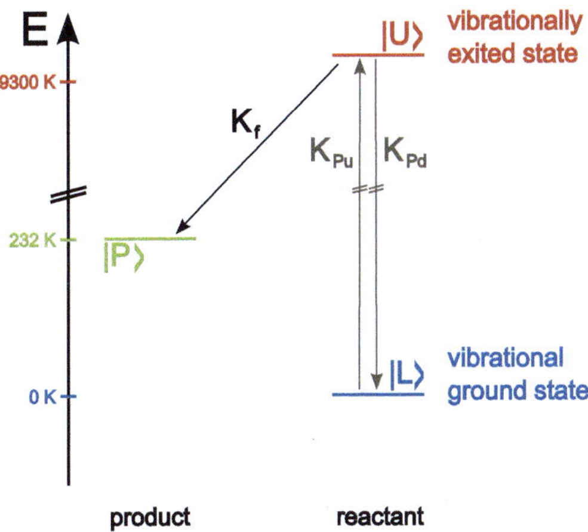

Figure 8.1: Simplified schematic of the energy levels and transition rates, that are needed for a numerical simulation of the LIR process in a reaction system similar to $H_3^+ + HD \rightleftharpoons H_2D^+ + H_2$. At least three levels and rates have to be taken into account:

Blue: A level $|L\rangle$ represents the vibrational ground state of the reactants as the lower state of the transition.

Red: A level $|U\rangle$ represents the vibrationally excited state of the reactants as the upper state of the pumped transition.

Green: A level $|P\rangle$ represents the products of the reaction.

Gray: The rates of laser absorption K_{Pu} and induced emission K_{Pd} of radiation transfer ions between the levels $|L\rangle$ and $|U\rangle$. (Spontaneous emission is neglected.)

Black: Excited ions colliding with the reaction gas form product ions with the reaction rate K_f.

with $B_{emis/abs}$: Einstein coefficients for induced emission and absorption,

$\qquad\qquad\qquad B_{abs} \cdot g_l = B_{emis} \cdot g_u$,

$\qquad g_{l/u}$: degeneracy of the lower/upper level of the transition,

$\qquad A_{emis}$: Einstein coefficient for spontaneous emission,

$\qquad p_{Laser}$: laser power,

$\qquad\quad \sigma$: standard frequency deviation of the Gaussian line profile given by the ion's velocity distribution as in (8.4),

$\qquad\quad \lambda$: wavelength of the pumped transition,

$\qquad\quad a$: illuminated area,

$\qquad\quad T$: temperature of the ions,

$\qquad\quad c$: speed of light,

$\qquad\quad h$: Planck constant,

$\qquad\quad k_B$: Boltzmann constant.

The ions are supposed to move quickly in the trap, so that every ion can be considered to cover the whole cross sectional area of the 22-pole trap on the timescales relevant here. Thus, taking the rather small cross section of the laser beam would underestimate a. Therefore, the cross-sectional area of the trap (assumed to be $\approx 5 \times 10^{-5}$ m^2) has to be used as illuminated area. By inserting typical experimental values ($p_{Laser} > 1$ W, $m \approx 4$ u, $T \approx 30$ K, $\lambda \approx 1546$ nm) it is possible to estimate the ratio of the rates for induced and spontaneous emission to be

$$\frac{K_{Pd}}{A_{emis}} \approx 100 \ . \tag{8.9}$$

Therefore, spontaneous emission can be neglected for the numerical simulations of this reaction system.

Reaction of Excited Ions to Products:

After reaching the upper reactant state $|U\rangle$, the ions can react in collisions with the reaction gas H$_2$ to the product state $|P\rangle$. The simplest approach would be to consider every collision of an excited ion with the reaction gas as reactive. In this case the rate K_f for the forward reaction of eq. (8.5) will be given by the number density of the reaction gas [H$_2$] and the Langevin rate coefficient k_L as:

$$K_f = [\text{H}_2] \cdot k_L \tag{8.10}$$

The Langevin rate coefficient describes the collision rate between charged particles and neutral gas without permanent dipole moment [66]:

$$k_L = 2.33 \times 10^{-9} \cdot q \sqrt{\frac{\alpha\left(\mathrm{\AA}^3\right)}{\mu\,(\mathrm{u})}} \tag{8.11}$$

$$= 1.79 \times 10^{-9}\,\frac{\mathrm{cm}^3}{\mathrm{s}} \tag{8.12}$$

q is the number of elementary charges on the ion (one in this case), $\alpha = 0.79\ \mathrm{\AA}^3$ the polarisability of the H_2 molecule (from [1]) and $\mu = \frac{4}{3}$ the reduced mass of the collisional system.

Frequency Dependence:

The ions in the trap have a Maxwell-Boltzmann velocity distribution given by their temperature. The Doppler shifts of the moving ions form a Gaussian line profile, when an absorption spectrum of such an ion cloud is measured. This can be included in the numerical simulation by assuming, that for any frequency only a reduced number of ions can be pumped to the upper state, following a Gaussian profile. For any time, first the number of ions in the lower state has to be calculated. Then this number has to be multiplied with a normalized Gaussian term.

This model will lead to different effective reaction rates for different frequencies (due to the Gaussian term). Thus, the ions with zero velocity in the direction of the laser beam will react faster than those absorbing Doppler shifted frequencies. When at a certain frequency all ions have reacted, the line will start to saturate. However, this model is oversimplified as it does not include any information on thermal level population.

8.2.2 Final Model for Numerical Simulations

Now we will include more internal states into the numerical model. This allows us to include physical informations like thermal level population and thermal repopulation of a level that is depleted by laser pumping. A schematic of this more complex model is shown in Figure 8.2 and will be explained in the following.

Introduction of further energy levels:

In reality the vibrational ground state of the ions contains several rotational levels J_{K_a,K_c} that are populated significantly at the typical experimental temperatures of some tens of Kelvins. For H_2D^+, most of the population will be in the four levels $0_{00}, 1_{01}, 1_{11}, 1_{10}$. For example at 30 K ion temperature, 99.72 % of all ions will be in one of those four levels. To be able to reproduce the thermal

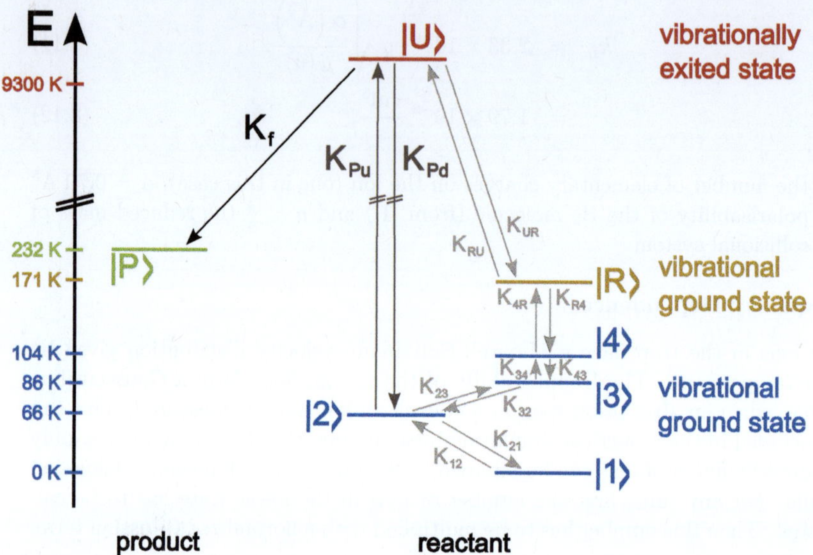

Figure 8.2: Schematic of the energy levels and repopulation rates between them, considered for the numerical simulations of the LIR process in a reaction system similar to $H_3^+ + HD \rightleftharpoons H_2D^+ + H_2$. For the simulation five levels were considered in the vibrational ground state, one in the vibrationally excited state of the reactants and the rovibrational ground state of the product ions. The lower transition level in this example is $|2\rangle$.

Blue: The energies are taken from the four lowest energy levels of H_2D^+.

Yellow: All remaining levels of the vibrational ground state are combined in the level $|R\rangle$ lying a little below the $J = 2$ levels in H_2D^+.

Red: For the vibrationally excited state of the reactants the upper state of the pumped transition was used.

Green: The products of the reaction are combined in just one level, whose energy does not enter the simulation. Given here is the endothermicity of the reaction, i.e. the difference between the lowest rovibrational states of reactants and products.

Black: Excited ions colliding with the reaction gas form product ions with the reaction rate K_f.

Grey: The rates of laser absorption K_{Pu} and induced emission K_{Pd} of radiation transfer ions between the levels $|L\rangle$ and $|U\rangle$.

Light Grey: The rates of thermal level population K_{ij} transfer ions from level $|i\rangle$ to level $|j\rangle$. To keep the model as simple as possible, only transitions between neighboring levels are considered in the simulation.

population of energy levels and the refilling of the lower transition level during the trapping process, all four levels with $J = 0, 1$ were included in the simulation. In Figure 8.2 these levels are shown in *blue*. For simplicity they are named $|1\rangle$ to $|4\rangle$ and the nuclear spin of the ions is neglected. Each of these four levels might be chosen as the lower level of the pumped transition. Figure 8.2 illustrates the situation, when level $|2\rangle$ is chosen as the pumped level. This option of choice makes it possible to investigate the influence of thermal population and of the Einstein B constant on the appearance of LIR lines.

Since the thermal population of the four lowest energy levels in H_2D^+ at 30 K contains only 99.72 % of the ions, another state was introduced which contains the remainig 0.28 % at 30 K. This state is thought to combine all other rotational levels of the vibrational ground state and is called $|R\rangle$. It is shown in *yellow* in Figure 8.2. The energy of this state is given by the assumption that it contains 0.28 % of the population at 30 K, which puts it at 171 K (a little below the $J = 2$ levels of H_2D^+ reaching from 189 K to 322 K).

In the simulation all levels in the vibrational ground state are considered to not react to product ions in collisions with the reaction gas. In reality, they have a very low probability to gain enough thermal energy for the reaction ($e^{-\frac{232\ K}{30\ K}} \approx 4 \times 10^{-4}$). Thereby they contribute to the constant background, but they do not influence the line profile.

Thermal Level Population Processes:

In addition to the transferring rates for laser pumping and the reaction rate from excited parent ions to product ions explained for the simple model above, also a set of transferring rates for the thermal population of energy levels has to be included. The thermal population of the levels forms in inelastic collisions of the ions with one of the gases present in the trap and by spontaneous emission of radiation. As explained above (Eq. (8.9)), the probability for spontaneous emission is so small, that it can well be neglected. Inelastic collisions with the reaction gas (H_2 in this case) or the buffer gas He redistribute the ions over the internal levels and change their velocity. The buffer gas is used to cool the ions down when they enter the trap, since they are produced at room temperature. For this purpose, the helium is injected to the trap via a pulsed valve at the beginning of the trapping process. This leads to a time dependence in the amount of He which would complicate the simulation a lot. Therefore, the Helium collisions are also neglected in this simulation and only collisions with H_2 are considered.

Usually the two different nuclear spin configurations (ortho and para) of H_2D^+ are treated as two different species, since conservation of nuclear spin strictly forbids spontaneous spin flips. Here these restrictions were not included, since collisions with H_2 are able to change the nuclear spin of H_2D^+.

Based on these assumptions, the easiest way to simulate thermal population processes is to introduce rates that transport an ion from its current energy level to the next higher or lower level in collisions with H_2, as shown in Figure 8.2 in *light grey*. These rates have to conserve the thermal population of the levels when no laser radiation is present. They are given by the collision rate multiplied by a factor giving the probability for a transfer of internal energy.

The rate K_{ij} from $|i\rangle$ to $|j\rangle$ is connected to the rate K_{ji} from $|j\rangle$ to $|i\rangle$ by a Boltzmann factor given by the energies $E_{i/j}$ of the levels and the temperature of the ions

$$K_{ij} = K_{ji} \cdot e^{\frac{E_i - E_j}{T_{ion}}} \,, \tag{8.13}$$

where the Boltzmann constant is included in the energies that are given in Kelvin. This conserves the thermal population without laser influence.

The rate for collisions with the reaction gas is again assumed to be given by the Langevin rate coefficient and the number density of H_2. The quenching probability is included by the factor q, which is considered a constant for all levels. With this, the rate from an upper level $|i\rangle$ to the next lower level $|j\rangle$ is given by:

$$K_{ij} = k_L \cdot q \cdot [H_2] \tag{8.14}$$

This rate is also considered for the upper level $|U\rangle$. Therefore, the reaction rate for the excited ions (8.10) has to be reduced by the quenching rate:

$$K_f = k_L \cdot (1 - q) \cdot [H_2] \tag{8.15}$$

In the simulations the quenching probability q is considered to be the same for all energy levels. In reality it will be given by the total number of levels with lower energy available in the product branch compared to the number of levels with lower energy available in the quenching branch.

So far, we considered only a redistribution of the internal states by the collisions with H_2. Of course, the velocity distribution will be influenced as well. This is included in the simulation by considering a Maxwell-Boltzmann distribution at any moment of the experiment. Therefore, the frequency dependence can be described in the same way as for the simple model explained above. The complete python [49] scripts for the numerical simulation are given in Appendix A.1.

Figures 8.3 to 8.5 show the time evolution of this numerical model for different parameters. Typical experimental values for the temperature, laser power and number density of the reaction gas were used in these simulations. Different values of q were used, to investigate the influence of the quenching probability on the line profiles.

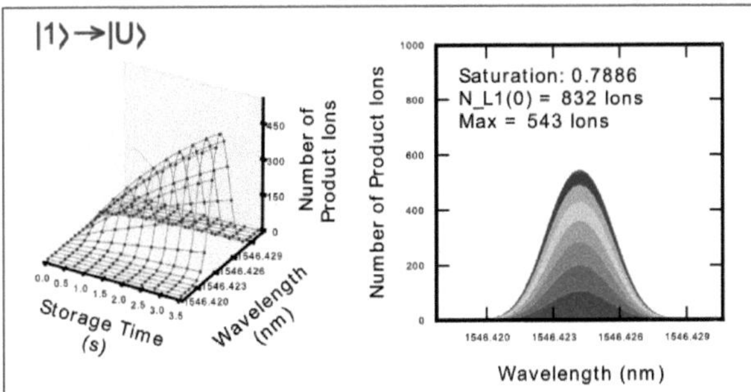

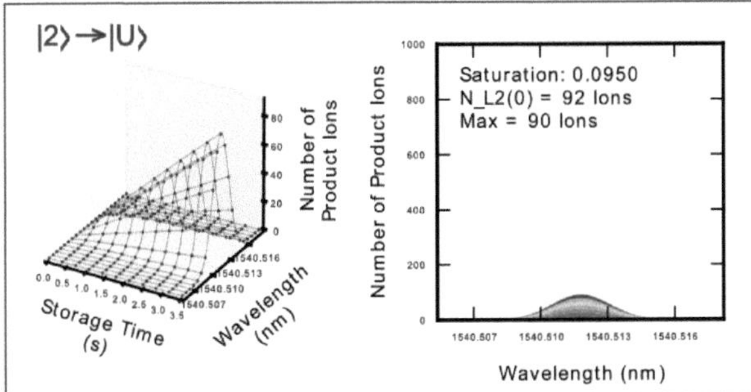

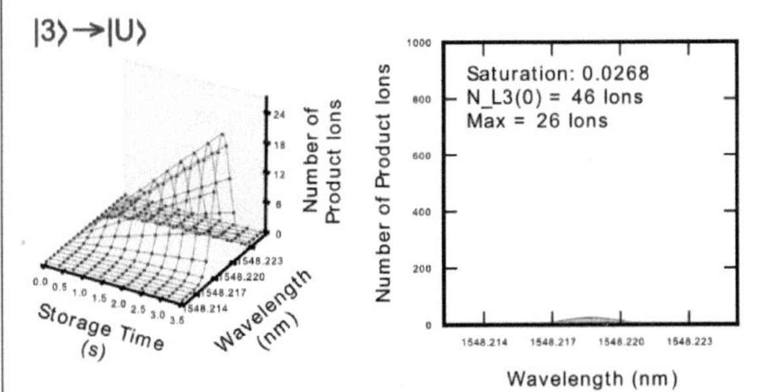

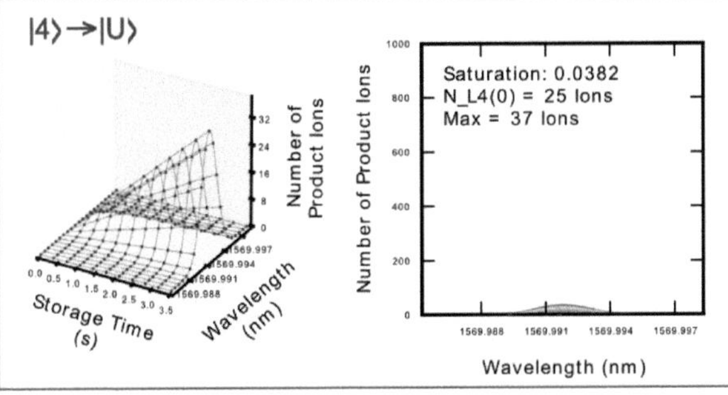

Figure 8.3:

Left:
Numerical simulation *(dots)* of the frequency dependent time evolution of the reaction system shown in Fig. 8.2 compared to function (8.3) *(wire frame)*.

Right:
Contour plot of function (8.3). All four plots have the same y scale for better comparability of the line strengths.

For each set of wire frame and contour plots, one of the levels $|1\rangle$ to $|4\rangle$ was chosen as ground state of the transition to compare the influence of thermal population and Einstein B constant on the line profile.

Simulation parameters:
$q = 0.8$
$T_{ion} = 30\ \mathrm{K}$
$p_{Laser} = 6\ \mathrm{mW}$
$[\mathrm{H_2}] = 7.5 \times 10^{11}\ \mathrm{cm^{-3}}$
total number of ions: 1000.

Parameters for the fit function:
$N_0 = 1000$,
$A \cdot t_{end}$ is given as *Saturation* in the contour plots of the different lines.
N_Li(0) gives the thermal population of level $|i\rangle$ at zero time,
Max is the number of product ions at the center frequency at the end of the reaction time.

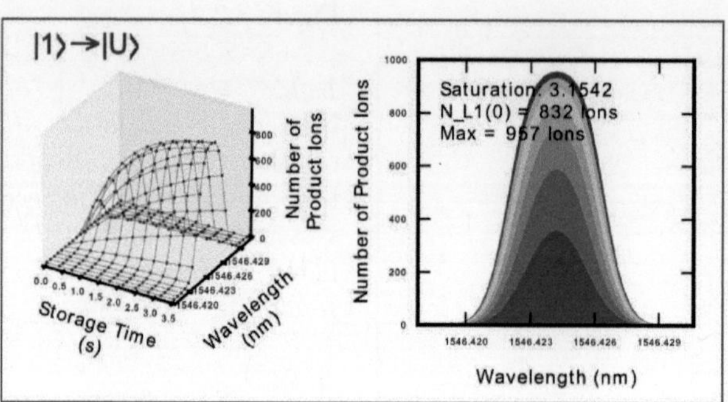

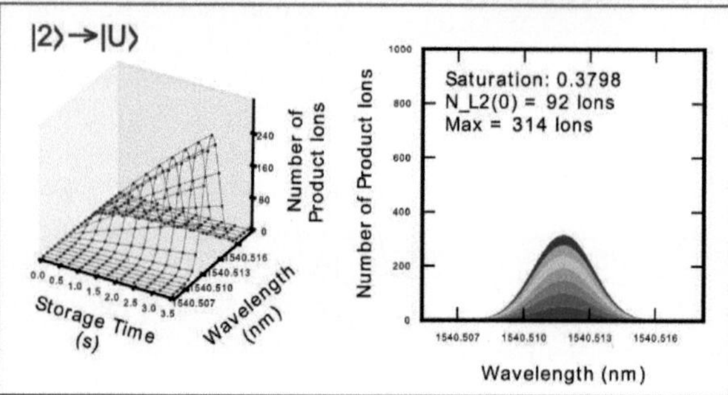

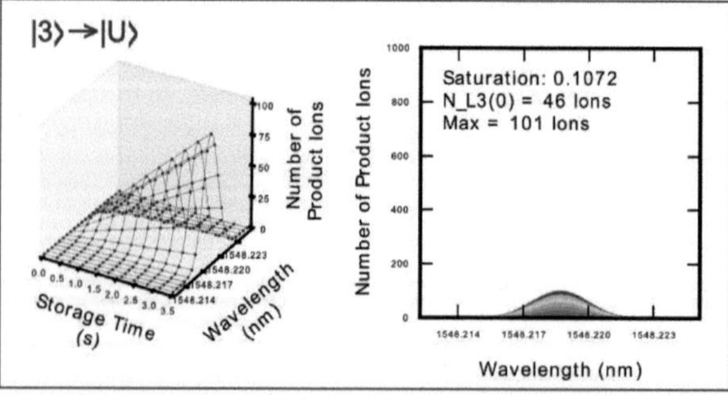

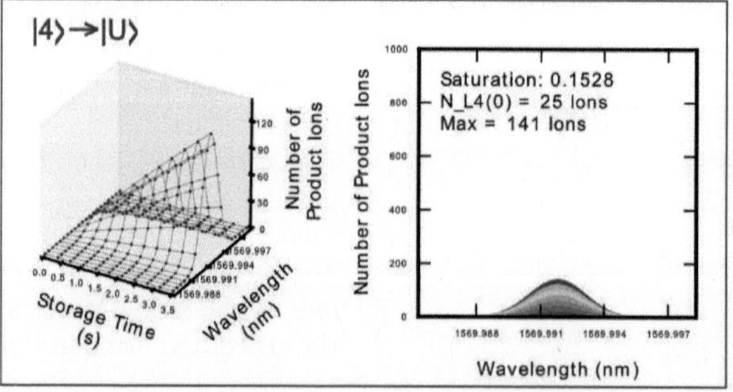

Figure 8.4:

Simulation parameters:
$q = 0.2$
$T_{ion} = 30\ \text{K}$
$p_{Laser} = 6\ \text{mW}$
$[\text{H}_2] = 7.5 \times 10^{11}\ \text{cm}^{-3}$
total number of ions: 1000.

For further details see Fig 8.3.

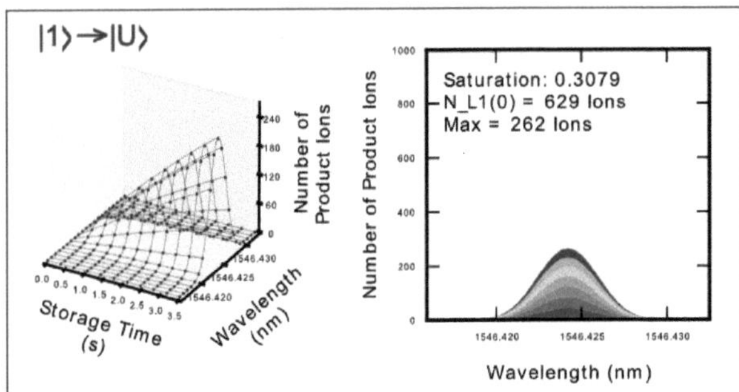

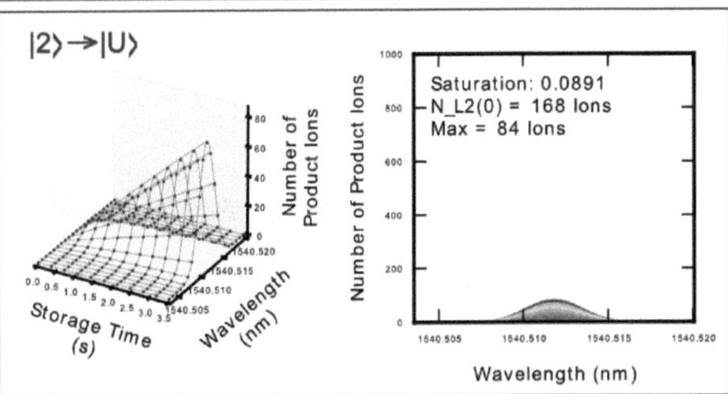

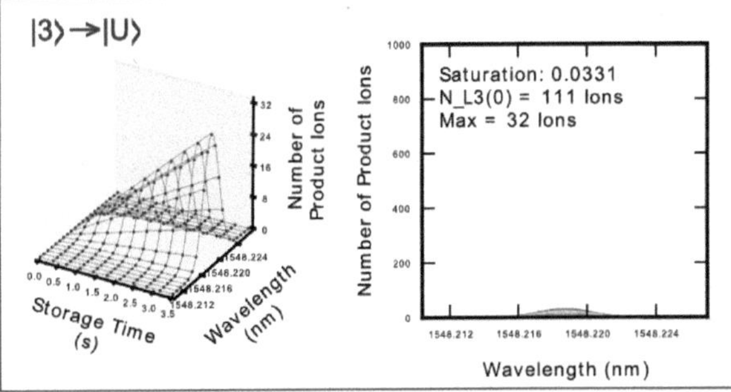

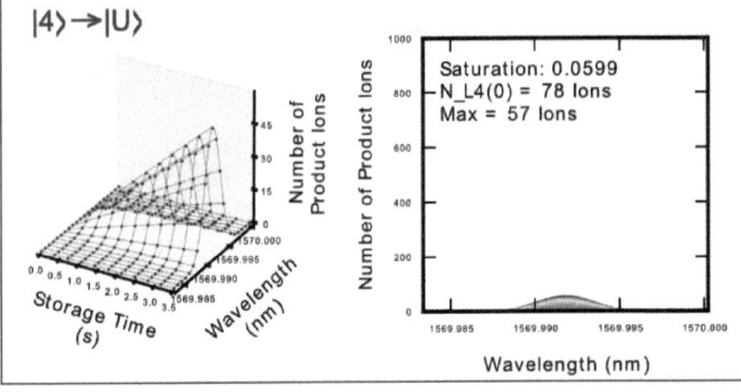

Figure 8.5:

Simulation parameters:
$q = 0.6$
$T_{ion} = 50$ K
$p_{Laser} = 2$ mW
$[H_2] = 5 \times 10^{11}$ cm^{-3}
total number of ions: 1000.

For further details see Fig 8.3.

Besides the numerical simulations, the figures include the wire frame plot of an analytic function of the form (8.3). It turned out, that the time dependent line profile of all numerical simulations can be described as

$$N_{P,i}(t,\lambda) \;=\; N_{total}\left(1 - e^{-A_i t e^{-\frac{mc^2}{2kT}\left(\frac{\lambda-\lambda_0}{\lambda_0}\right)^2}}\right) , \tag{8.16}$$

$$A_i \;=\; 0.2 \cdot (1-q) K_{Pu}\frac{N_{Li}(0)}{N_{total}} , \tag{8.17}$$

where $N_{Li}(0)$ is the number of ions in level $|i\rangle$ at $t = 0$, N_{total} is the sum of ions in all levels, K_{Pu} (see Equation (8.8)) and q are defined as given above. The factor of 0.2 has been introduced to get the best match between numerical simulation and analytical function, it has no physical motivation and is not understood so far. A_i includes the normalization factor of the Gaussian line profile $\frac{1}{\sigma\sqrt{2\pi}}$ and thus the temperature of the ions.

The numerical simulations described here, were able to reproduce the time and frequency dependence known from LIR processes qualitatively. This is remarkable, since we still included the frequency dependence as a Gaussian profile for any time of the simulations. The saturation of the lines is formed by the repopulation processes of the non reacting internal states. The saturated lines produced by the simulations can be described by a function of the theoretically derived form (8.3), for all sets of simulation parameters. The simulations give first hints which parameters might influence the observed saturation of measured lines and in which way. Figures 8.3 to 8.5 show that the Einstein B coefficient and the energy of the lower state of a transition as well as the quenching probability of the upper state and the ion temperature strongly influence the observed line profiles.

Since the reaction system for a numerical simulation has to be strongly simplified in comparison to reality, it was not possible to apply these simulations to reproduce LIR measurements of CH_2D^+ taken earlier (Chapter 7). As these experiments were originally intended to measure accurate line positions, some relevant parameters which are needed to understand the line profiles quantitatively were not measured (e.g. the laser power or the number of parent ions inserted in the trap). Therefore, new measurements had to be performed, where all possibly relevant parameters were recorded. Those were performed on CH_5^+ and are presented in the following.

8.3 Test Measurements on CH_5^+

A series of measurements was performed to investigate the relations between the fit parameters used for saturated lines and the conditions chosen for the LIR experiments. The fit function for saturated lines includes the following parameters: line strength (i.e. the number of parent ions that are available for the reaction, N_0), line width (i.e. standard deviation of the Gaussian term), saturation (i.e. $A \cdot t$), and background (i.e. $C(\nu)$). Experimental parameters that are expected to influence the profile of the measured LIR lines are: number of parent ions inserted in the trap, trap temperature, laser power, reaction time, and Einstein B coefficient of the investigated transition. The systematic investigation of the effects of these parameters is presented in the following. The results are summarized in Table 8.1, the colour coding of Figures 8.11 to 8.19 is also provided in this table. The main goal of these measurements was to verify the expected time-dependence of the LIR signal and to get insight in the thermal repopulation processes of the internal states of the reactant ions. The latter is necessary to deduce the thermal level population and thus the rotational temperature from the observed lines in future experiments.

The required measurements on the influence of experimental conditions on the fit parameters were performed on the endothermic reaction system

$$CH_5^+ + CO_2 \rightleftharpoons CO_2H^+ + CH_4 + 361 \text{ K} , \qquad (8.18)$$

which was investigated in the 22-pole ion trap experiment before [2]. It is not the ideal system for such measurements, as will be explained in the following. However, due to recent progress in the frequency resolution (availability of an OPO and a frequency comb) the spectroscopy on CH_5^+ was revived [67]. As the change to other reaction systems and radiation sources usually takes several weeks on LIRTrap, the measurements presented here, were performed on CH_5^+.

8.3.1 Experiments on CH_5^+

The experiments with CH_5^+ require some modifications to the LIR experiments described before. As the reaction gas CO_2 freezes out at the cryogenic temperatures used for the experiments, it was inserted in the trap via a second pulsed valve in a mixture with He. Therefore, the number density of CO_2 was not constant over the reaction time. This excludes measurements of the temporal evolution of the LIR signal, because the reaction rate K is given by a rate coefficient k multiplied with the number density of the reaction gas. A time dependent rate would complicate things too much.

Another change is that the storage time and the laser irradiation time are not the same anymore. In order to reduce the background signal on the product mass CO_2H^+ and the width of the observed lines, the laser was blocked by a shutter for the first 30 ms of the storage time. At that time, also no CO_2 was present in the trap. This ensures cooling of the ions to the ambient cryogenic temperature before the LIR process started.

For the measurements one of the strongest transitions (at 10 K) of CH_5^+ was chosen. It is located at 3041.6355 ± 0.0012 cm^{-1}. This line slightly overlaps[3] with a weak transition at 3041.6307 ± 0.0017 cm^{-1}.

8.3.2   OPO System

A commercially available OPO[4] was used as radiation source. For precise frequency determination, a frequency comb[5] in combination with a wavemeter[6] and a rubidium clock[7] was used. The setup is explained in more detail in [67]. A brief explanation of the OPO principle is given in Chapter 7.2.2 on the example of the homebuilt OPO.

8.3.3   Fitting the Data

Figure 8.6 shows three examples of the measurements on CH_5^+. For most measurements several scans of the line at 3041.6355 cm^{-1} were taken to improve the signal to noise ratio. There are often slight drifts in the experimental conditions over the measurement time. To correct for that, the data were scaled to get the best possible overlap between the different scans of the line without changing the average values of background height and line strength for the measurement. Afterwards data points with the same frequency were averaged.

All data were fitted with a least-squares routine in `python` [49], using the function:

$$f(x) = N_0 \cdot \left(1 - e^{-\frac{a}{\sigma\sqrt{2\pi}} e^{-\frac{(x-x_0)^2}{2\sigma^2}}} \right) + N_{0,1} \cdot \left(1 - e^{-\frac{a_1}{\sigma_1\sqrt{2\pi}} e^{-\frac{\left(x-x_{0,1}\right)^2}{2\sigma_1^2}}} \right) + C + C_2 \cdot (x - x_{0,1})$$

$$(8.19)$$

N_0 is the number of ions available for the reaction and is called line strength in the following, x_0 is the position of the line, a its saturation[8] and σ the standard

---

[3]The overlap became obvious during the measurements with higher nominal temperatures.

[4]Aculight Argos Model 2400, $2560 - 3130$ cm^{-1}

[5]Menlo System FC1500

[6]Bristol Instruments Model 621 A-IR

[7]Rhode&Schwarz XSRM

[8]Since the temperature of the ions influences the linewidth and thus the dimensionless parameter $A \cdot t_{end}$, we now use the parameter a instead $\left(\frac{a}{\sigma\sqrt{2\pi}} = A \cdot t_{end} \right)$, which has the same dimension like the line width, cm^{-1} in this case.

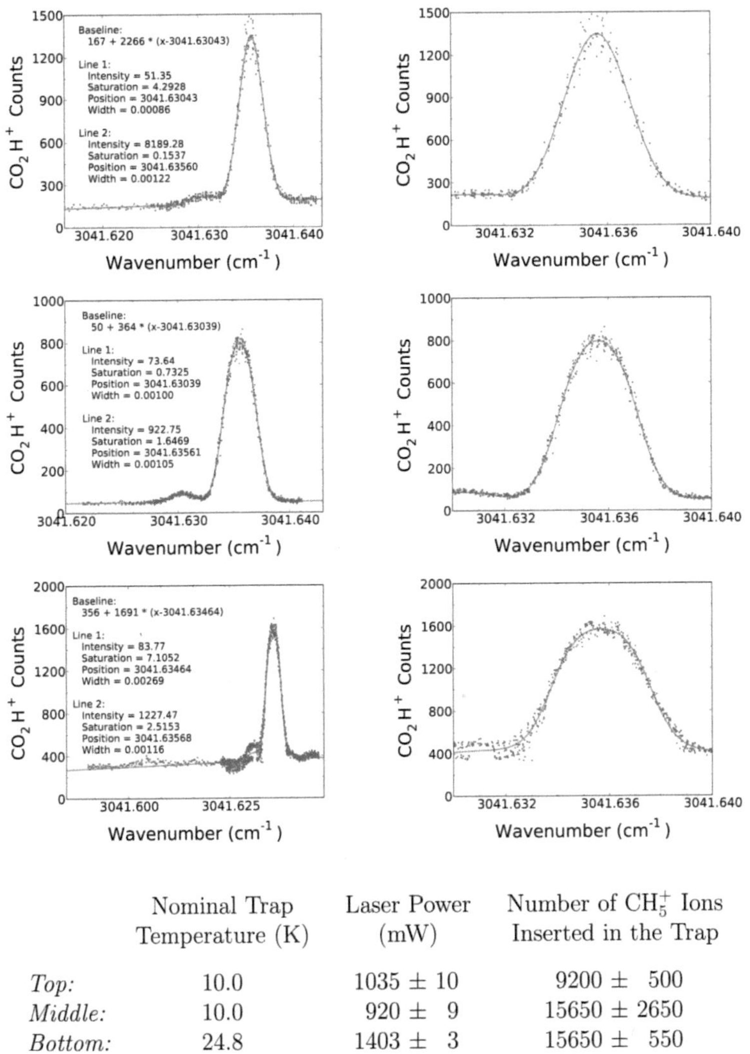

	Nominal Trap Temperature (K)	Laser Power (mW)	Number of CH_5^+ Ions Inserted in the Trap
Top:	10.0	1035 ± 10	9200 ± 500
Middle:	10.0	920 ± 9	15650 ± 2650
Bottom:	24.8	1403 ± 3	15650 ± 550

Figure 8.6: Examples of line profiles measured on CH_5^+. The CH_5^+ ions were trapped for 920 ms. For the first 30 ms of the storage time, the laser was blocked by a shutter.

Left: Data points with the same frequency were averaged and function (8.19) was fitted.

Right: Enlarged view of the stronger line.

The fit results are given in the left graph. Line 1 is the weaker, line 2 the stronger line.

deviation of the Gaussian term - called line width in the following. The same notation is used for the parameters fitting the weaker line with the index $_1$ added. The baseline is described by the last two terms with C the baseline height at the center of the weak line (approximately the center of the scanned frequency range) and C_2 the slope of the baseline.

In the next sections, we will investigate the influence of the following experimental parameters: number of parent ions inserted to the trap, laser power, reaction time, and trap temperature. All lines that are shown in the following plots are thought as guide to the eye, to show what kind of trend is expected. They are not fits of the data, as the quality of the measurements is not sufficient to deduce quantitative correlations.

8.3.4   Influence of the Number of Parent Ions

Three series of measurements were taken to investigate the effect of the number of parent ions on the line profile. They were all performed at a nominal trap temperature of 10 K and a storage time of the ions of 920 ms. The shutter blocking the laser opened after the first 30 ms, i.e. the laser irradiated the ions for the remaining 890 ms. Three measurement series with different laser powers of 920 mW, 1035 mW, and 1180 mW, respectively have been recorded. The data were fitted with function (8.19).

Line Strength:

One of the fit parameters in function (8.3) is the line strength N_0. It is thought to be the number of ions available for the reaction. This might be all the parent ions inserted in the trap or just some fraction of this number, given by the thermal distribution of the ions over the internal states or by the geometry of the trapping fields. Whatever effect might reduce the number of available ions w.r.t. the total number inserted, both parameters are expected to be proportional.

The results of this investigation are shown in Figure 8.7. The graph shows the expected trend, though the quality of the measurements is not sufficient to derive the proportionality factor. The plot indicates, that the proportionality factor is well below one. Thus, only a small fraction of the ions inserted in the trap take part in the reaction. As indicated by the solid lines in Figure 8.7, the proportionality factor seems to be dependent on the laser power. We would not have expected this correlation from the considerations in Section 8.1.

There are several possible explanations for a reduced number of reacting ions w.r.t the inserted number.

1. The number of parent ions that travel through the laser beam might be reduced. Such effects can be caused by imperfections in the trapping poten-

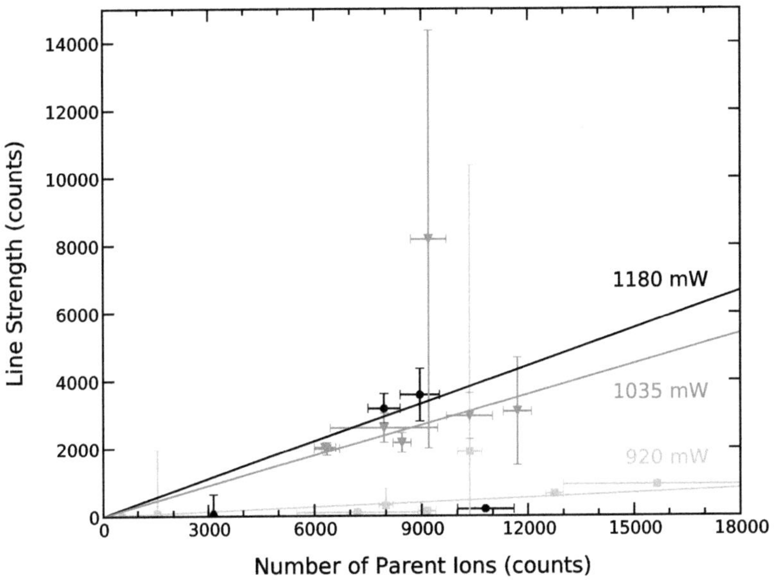

Figure 8.7: Dependence of the line strength N_0 on the number of parent ions inserted in the trap. Three measurement series at different laser powers (1180, 1035, and 920 mW) are shown. The solid lines are not fits but serve as guide to the eye for the expected trend.

tial. The electric field trapping the ions might create not only one minimum of the trapping potential, but several.

2. The thermal population of the pumped level and its refilling probability via inelastic collisions might be rather low. This could behave like a reaction system with a reduced number of parent ions in the experiments.

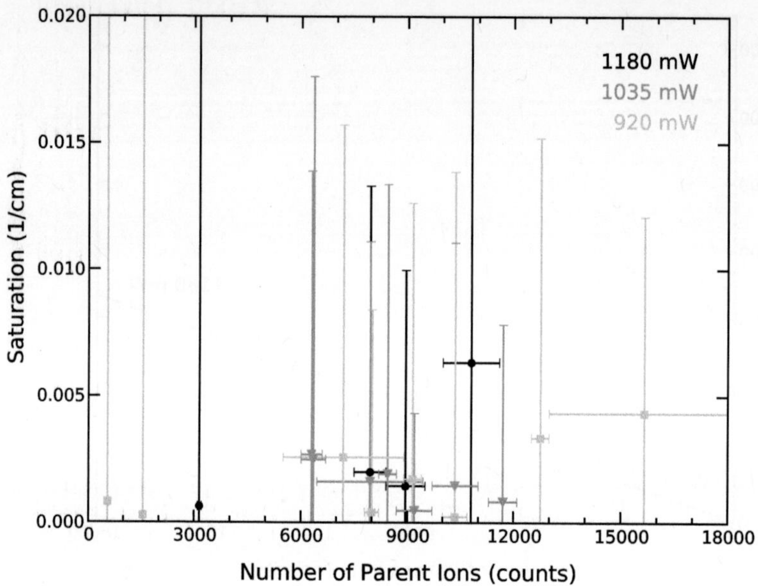

Figure 8.8: Dependence of saturation a on the number of parent ions inserted in the trap. Three measurement series at different laser powers (1180, 1035, and 920 mW) are shown.

Saturation:

The saturation of a line is supposed to include only the effective reaction rate for the induced reaction and the time, the reaction had to evolve. Thus, it should be independent of the number of parent ions. Figure 8.8 shows the saturation of the spectral line as a function of the number of parent ions inserted in the trap. The observed saturation values are constant within the fit uncertainties. Nevertheless, some care has to be taken with the interpretation of the results from CH_5^+ measurements. Comparison of the data points in Figures 8.7 and 8.8 shows, that all fits leading to very high line strengths showed very low saturations and vice versa. A possible explanation for this correlation is, that both fit parameters, line strength and saturation influence the maximum height of the line profile. Hence, for low saturations (i.e. only slight deviations from the Gaussian line profile) they are not completely independent in the fit. To make the fits more reliable, experimental conditions leading to higher saturations should be used for future measurements.

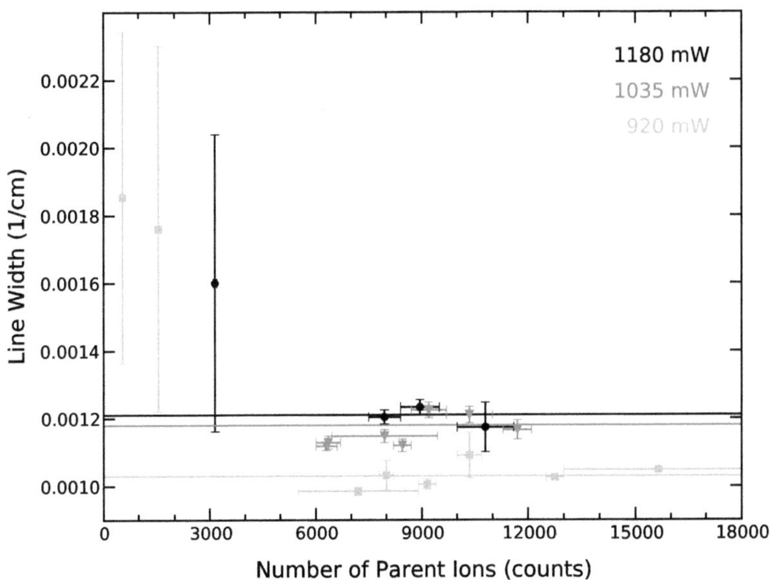

Figure 8.9: Dependence of the line width σ on the number of parent ions inserted in the trap. Three measurement series at different laser powers (1180, 1035, and 920 mW) are shown. The solid lines are not fits but serve as guide to the eye for the expected trend.

Line Width:

Figure 8.9 shows a plot of the observed line width as a function of the number of parent ions. The theory described above predicts the line width to be independent on the number of parent ions. For most of the data points within one series this seems to be true. Comparing the measurement series with different laser powers, there appears to be a slight increase of the line width towards higher laser powers. Some care has to be taken here as well, since the fit of the line width is not completely independent from the fit of the saturation - both parameters make lines broader. The Doppler width expected for a temperature of 10 K is 0.0007 cm^{-1}.

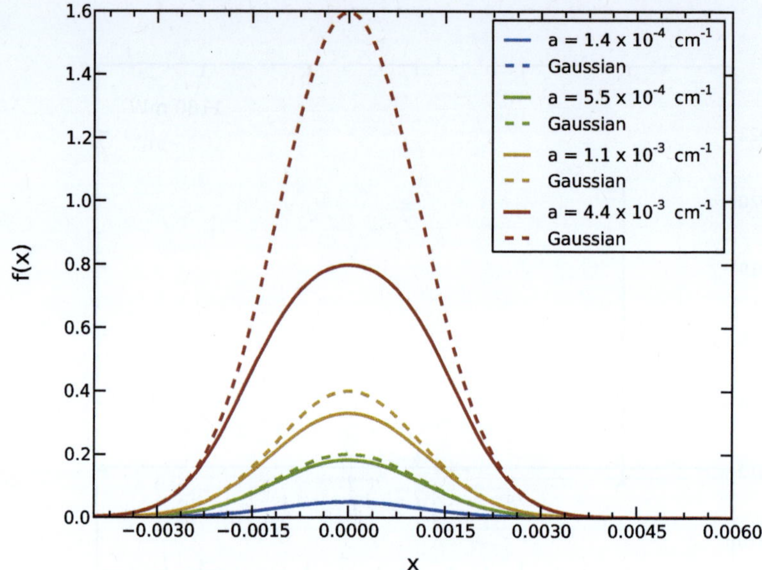

Figure 8.10: Saturated line profiles in comparison with Gaussian profiles for different values of the saturation a (cm^{-1}). Parameters: $\sigma = 1.1 \times 10^{-3}$ cm^{-1}, $N_0 = 1$.

Solid lines: Saturated profiles, $f(x) = 1 - \exp(-\frac{a}{\sigma\sqrt{2\pi}}\exp(-\frac{x^2}{2\sigma^2}))$.

Dashed lines: Gaussian profiles, $f(x) = \frac{a}{\sigma\sqrt{2\pi}}\exp(-\frac{x^2}{2\sigma^2})$.

Correlation of Fit Parameters for Low Saturations:

Figure 8.10 shows saturated line profiles (Eq. (8.3)) in comparison with Gaussian functions for different values of the saturation $a = At\sigma\sqrt{2\pi}$. The line strength N_0 was set to one for all curves. The line width was chosen as $\sigma = 1.1 \times 10^{-3}$ cm^{-1}, which is comparable to the experimental observations.

Obviously, the deviation from a Gaussian line profile becomes clear only for saturations larger than $a \approx 10^{-3}$ cm^{-1}. The blue and green solid lines in Figure 8.10 can be reproduced easily by a Gaussian term with slightly changed values for line width and line strength compared to the dashed curve of the same colour. Depending on the quality of the measured data, the confusion between line strength, saturation and width can very well hold up to saturations of $a \approx 1.5 \times 10^{-3}$ cm^{-1}.

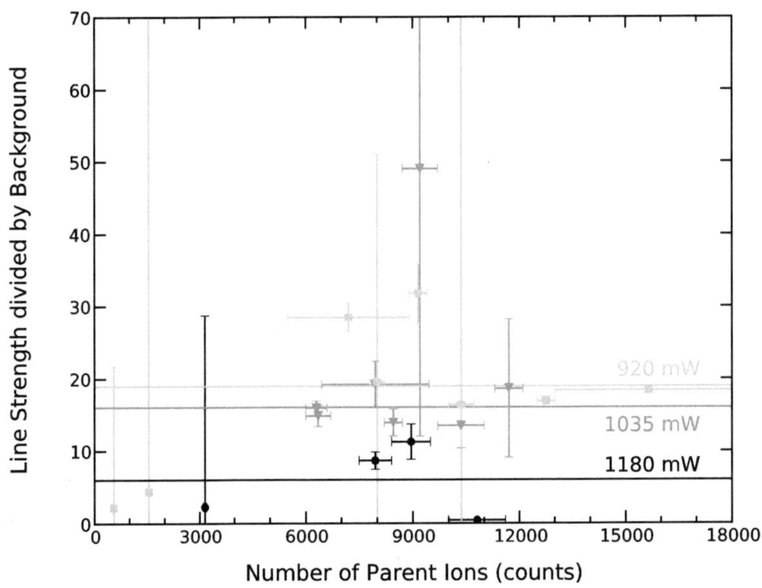

Figure 8.11: Dependence of the ratio line strength to background $\left(\frac{N_0}{C}\right)$ on the number of parent ions inserted in the trap. Three measurement series at different laser powers (1180, 1035, and 920 mW) are shown. The solid lines are not fits but serve as guide to the eye for the expected trend.

Line Strength to Background Ratio:

The amount of hot ions reacting without laser irradiation will most likely be proportional to the number of parent ions. Thus, the ratio $\frac{\text{line strength}}{\text{background}}$ should be independent of the number of parent ions and might serve to investigate the influence of other experimental parameters, as it is not always possible to keep the number of parent ions constant over a measurement series. Figure 8.11 shows the results of this correlation. For the measurement series at 1035 mW the values for the ratio $\frac{\text{line strength}}{\text{background}}$ are constant within the given fit uncertainties, for the two other series, they are not. This is most likely due to the confusion between saturation and line strength as explained above.

8.3.5   Influence of the Laser Power

Three series of measurements were taken to investigate the dependence of the
fit parameters on the laser power. For all measurements the storage time was
920 ms, with the first 30 ms without laser irradiation. Two series were performed
at a nominal trap temperature of 10.0 K, one at 24.8 K. For the first series at
10.0 K, the number of parent ions increased from 6050 to 13550 from the first to
the last measurement. For the other two series, it varied much less (from 13900
to 11200 in the second series at 10.0 K and between 14400 and 16450 for the
series at 24.8 K). The data were fitted with function (8.19).

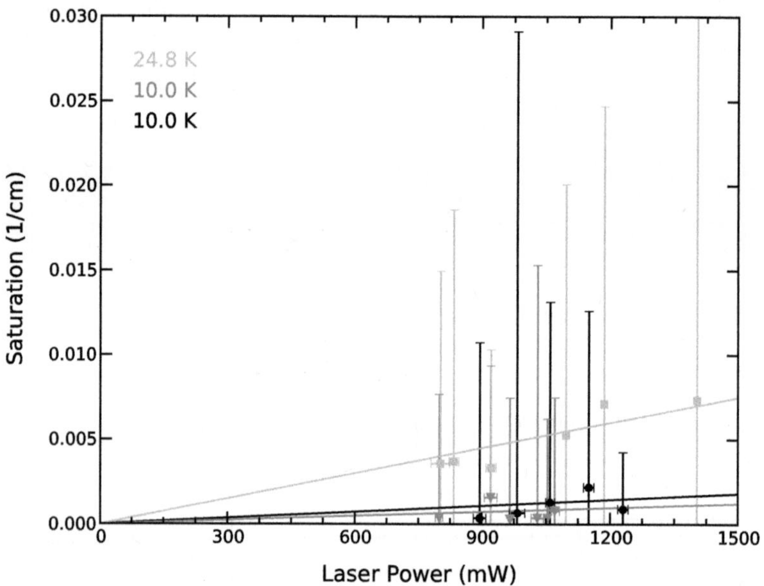

Figure 8.12: Dependence of the saturation a on the laser power. Three measurement series at two different nominal trap temperatures (10.0 and 24.8 K) are
shown. The solid lines are not fits but serve as guide to the eye for the expected
trend.

Saturation:

As explained above, we would expect the saturation to be the only fit parameter
that depends on the laser power. The effective rate for the endothermic reaction A will be proportional to the number of ions absorbing a photon. Thus, it

has to be proportional to the number of photons available for absorption. Therefore, we expect a linear dependence of the saturation $a = At\sigma\sqrt{2\pi}$ on the laser power. The experimental results for this dependence are shown in Figure 8.12. Each measurement series shows the expected linear behaviour within the fit uncertainties. As explained above, the fits for saturations below $a \approx 1 \times 10^{-3}$ cm^{-1} or sometimes even 1.5×10^{-3} cm^{-1} tend to show confusion between the fit parameters saturation, line strength and line width. For this reason, the measurement series at 24.8 K is far more reliable than the other two.

Line Strength:

The line strength[9] of the observed lines is expected to be independent from the chosen laser power. We expect it to depend on the number of parent ions inserted in the trap, which varied for the measurements presented in this section. This effect can be reduced by taking the ratio $\frac{\text{line strength}}{\text{background}}$ instead of the line strength, as both background and line strength are expected to be proportional to N_0. The plot line strength versus laser power is given in Figure A.1 in the Appendix A.2 for completeness.

---

[9]Line strength means the number of parent ions available for the reaction, N_0 in Equation (8.19).

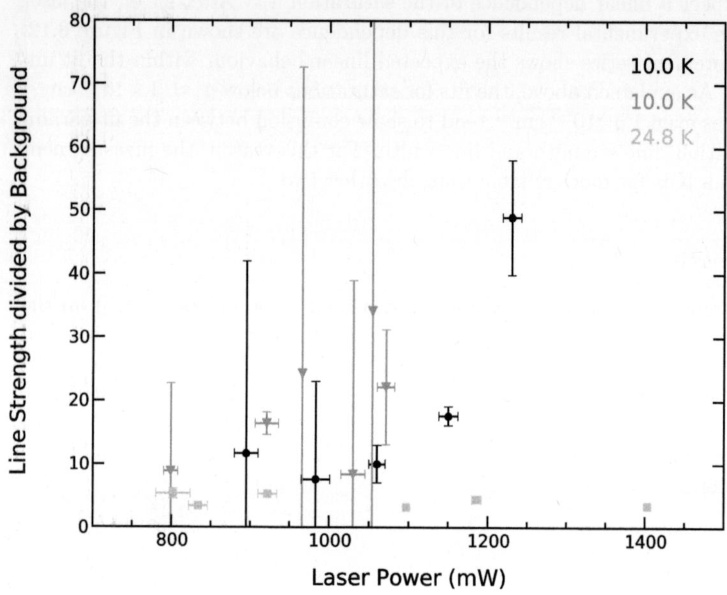

Figure 8.13: Dependence of the ratio line strength to background $\left(\frac{N_0}{C}\right)$ on the laser power. Three measurement series at two different nominal trap temperatures (10.0 and 24.8 K) are shown.

Line Strength to Background Ratio:

The background counts of the measured lines evolve without laser radiation. They are formed by parent ions that thermally gained enough energy for the reaction. Thus, they have to be independent from the laser power. They can serve to correct for experimental drifts in the number of parent ions inserted in the trap. As explained above, the ratio $\frac{\text{line strength}}{\text{background}}$ should thus be independent from the laser power. The measurement series at 10.0 K are inconclusive on that point (see Figure 8.13), they seem to be increasing with increasing laser power. This effect is probably due to the correlation between saturation and line strength in the fits for low saturation values (compare Figure 8.12). The data points for the series at 24.8 K scatter less than in Figure A.1, but they do not agree within the errors.

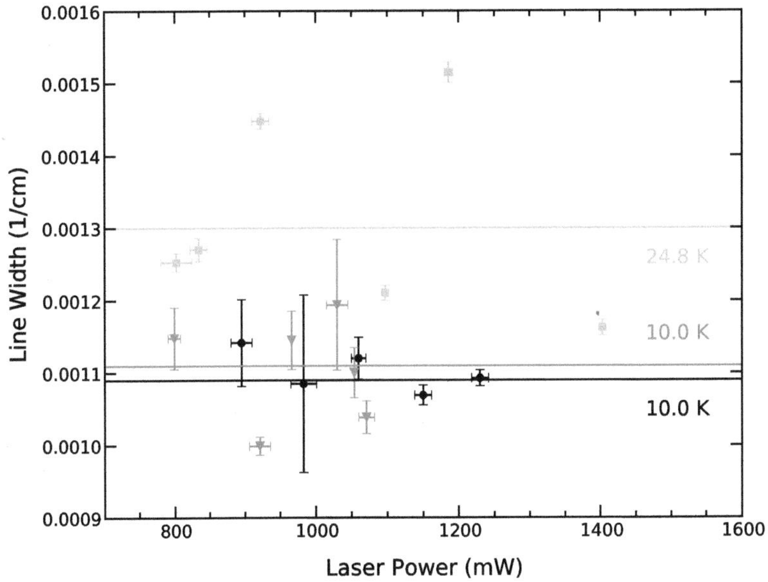

Figure 8.14: Dependence of the line width σ on the laser power. Three measurement series at two different nominal trap temperatures (10.0 and 24.8 K) are shown. The solid lines are not fits but serve as guide to the eye for the expected trend.

Line Width:

We expect no effect of the laser power on the line width. The results (Figure 8.14) do not match within the error bars, but neither a trend is showing. As expected, the line widths of the measurements at 24.8 K are higher than those at 10.0 K. The Doppler width expected for a temperature of 10.0 K is 0.0007 cm^{-1}, for 24.8 K we expect 0.0011 cm^{-1}.

8.3.6   Influence of the Time

Two series of measurements were performed to investigate the dependence of the different fit parameters on the irradiation and the storage time. In one series the storage time was varied between 220 and 420 ms while the laser irradiation time was kept at 320 ms (from 30 to 350 ms after entrance of the ions). In the other series the storage time was kept constant at 920 ms while the laser irradiation time was varied between 400 and 900 ms (starting always 30 ms after entrance of the ions). The data were fitted with function (8.19).

At the beginning of the measurement series, the laser power was monitored. It was 1163 mW for the first series and 1020 mW for the second series. The number of parent ions varied between 7100 and 14600 for both measurement series. As explained above, this might influence the line strength, so some care has to be taken with the interpretation of Figure 8.17.

The y-errors for the saturation and line strength plots are especially large for the measurements presented in this section. This is due to the rather low values for the saturation and the already explained confusion between saturation and line strength. As the saturation is expected to grow with time, the correlation of the two fit parameters is likely to become stronger towards shorter times. The y-errors in the graphs are the uncertainties of the `python` [49] fit results, while the x-errors represent experimental drifts and/or accuracies.

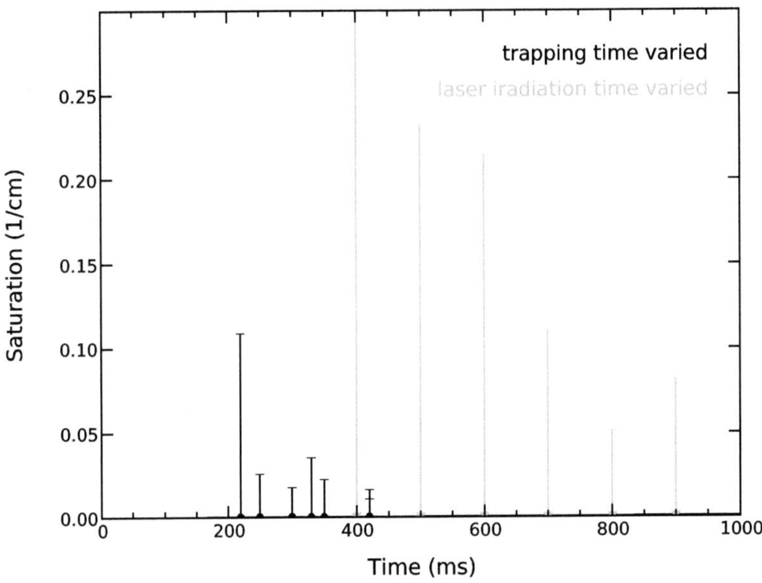

Figure 8.15: Dependence of the saturation a on the irradiation and the storage time. Two measurement series are shown, for one series the storage time of the ions was varied, for the other the laser irradiation time. The solid lines are not fits but serve as guide to the eye for the expected trend.

Saturation:

We would expect that the saturation increases linearly with time as it is thought to be $a = At\sigma\sqrt{2\pi}$, with the effective rate for the endothermic reaction A. The results for this dependence are shown in Figure 8.15. Due to the low quality of these measurement series, the saturation is constant within the error bars. The results seem to scatter around a linear increase (see enlarged view in Figure A.2 in the Appendix A.2) for both measurement series. This is a promising start, but measurements have to be repeated. There are two possible improvements for future measurements: Higher saturations would help to reduce the confusion between the fit parameters and a constant number of parent ions for all measurements would allow to fix the line strength fit parameter. Fixing N_0 would allow to assure, that the observed time-dependence of the saturation is not an artifact of a fit routine.

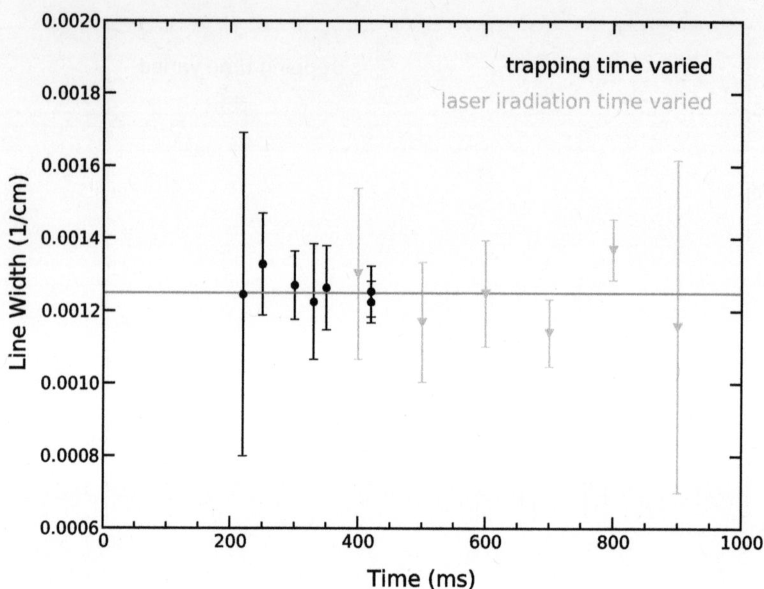

Figure 8.16: Dependence of the line width σ on the irradiation and the storage time. Two measurement series are shown, for one series the storage time of the ions was varied, for the other the laser irradiation time. The solid line is not a fit but serves as guide to the eye for the expected trend.

Line Width:

The line width σ is expected to be time-independent. This is confirmed by Figure 8.16.

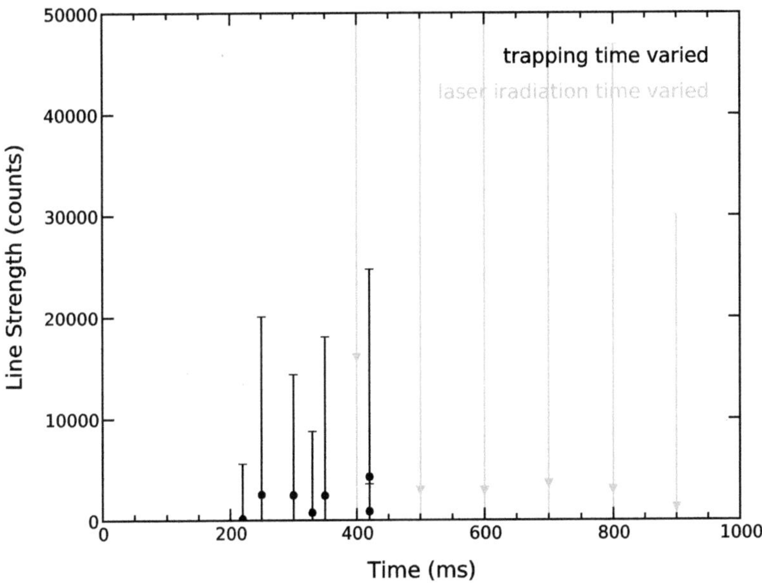

Figure 8.17: Dependence of the line strength N_0 on the irradiation and the storage time. Two measurement series are shown, for one series the storage time of the ions was varied, for the other the laser irradiation time.

Line Strength:

The line strength N_0 also should be time-independent. In Figure 8.17 all values are constant within the error bars. As discussed above the errors are very large, since low saturations lead to correlations between the fit parameters saturation, line strength, and line width. Also, the line strength is most likely proportional to the number of inserted parent ions, which varied for the measurements presented here. Usually, in this case, the plot of the $\frac{\text{line strength}}{\text{background}}$ ratio can be helpful. For the time-dependence this might not be case as discussed below.

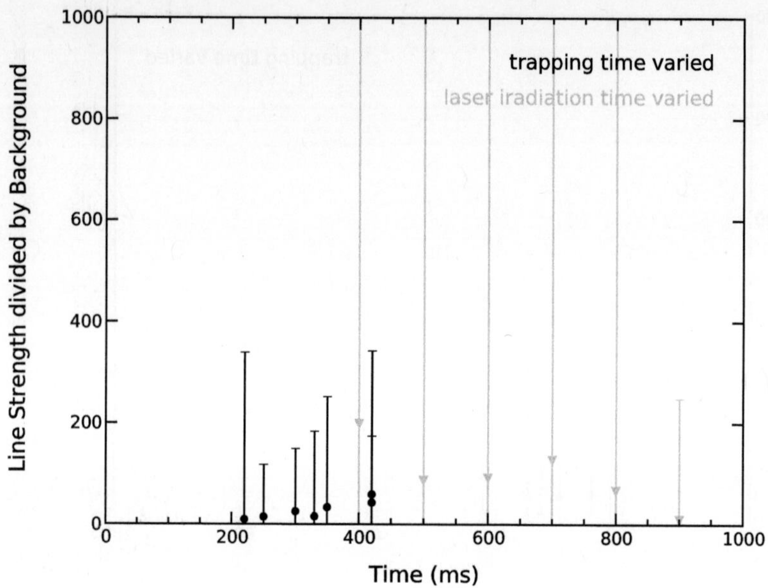

Figure 8.18: Dependence of the ratio line strength to background $\left(\frac{N_0}{C}\right)$ on the irradiation and the storage time. Two measurement series are shown, for one series the storage time of the ions was varied, for the other the laser irradiation time.

Line Strength to Background Ratio:

The expectancy for the ratio $\frac{\text{line strength}}{\text{background}}$ is not clear so far. The background might increase with time, which would lead to a decrease in $\frac{N_0}{C}$. The data in Figure 8.18 show no clear trend, but again the quality of these measurements was not sufficient to obtain definite results.

8.3.7 Influence of the Trap Temperature

Two series of measurements were performed to investigate the influence of the trap temperature on the fit parameters. The timings were again chosen to be 920 ms storage time with laser radiation present for the last 890 ms. Preferably, the number of parent ions CH_5^+ should have been constant for this measurement series. This was not possible, as the pressure of the precursor gas for the ion production usually drifts slowly. Sometimes also other experimental parameters that influence the number of trapped ions like the intensity of the helium pulse or some trapping potentials show such drifts. For the measurements presented here, the number of ions inserted to the 22-pole trap varied between 6000 and 11850 for the first series and between 12750 and 23350 for the second series. The laser power for the first series was stable at 1200 mW, for the second series it varied between 1173 and 1280 mW. The data were fitted with function (8.19).

The nominal trap temperature is the value read from the silicon diode mounted on the inner trap chamber. The ions are usually a few Kelvin warmer. Their temperatures can be derived from the Doppler widths of the observed lines, when these lines show a Gaussian profile. According to the theories about line saturation discussed so far, we expect to be able to derive translational ion temperatures from the Gaussian profile inserted into Lambert-Beer's law. Since the investigations presented here are meant to verify this assumption, we cannot use this Doppler width to determine the ion temperature, but we have to take the nominal trap temperature instead.

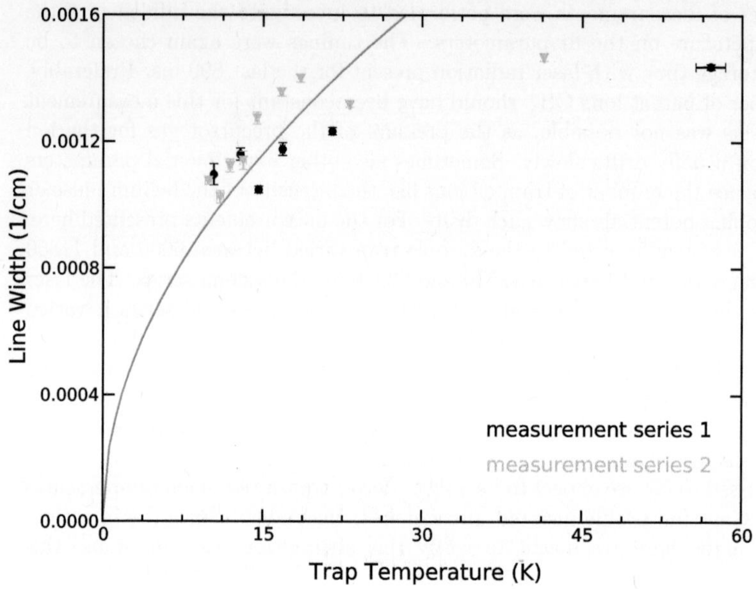

Figure 8.19: Dependence of the line width σ on the trap temperature. Two measurement series at similar experimental conditions are shown. The solid line is not a fit but serves as guide to the eye for the expected trend.

Line Width:

The standard deviation of the Gaussian term included in the fit function depends on the ion temperature as $\sigma = \nu_0 \sqrt{\frac{kT_{ion}}{mc^2}}$. The dependence of the observed line width on the trap temperature should therefore follow a $\sqrt{T_{trap}}$ dependence. The trend of the results shown in Figure 8.19 appears to follow that law, although they do not match within the error bars.

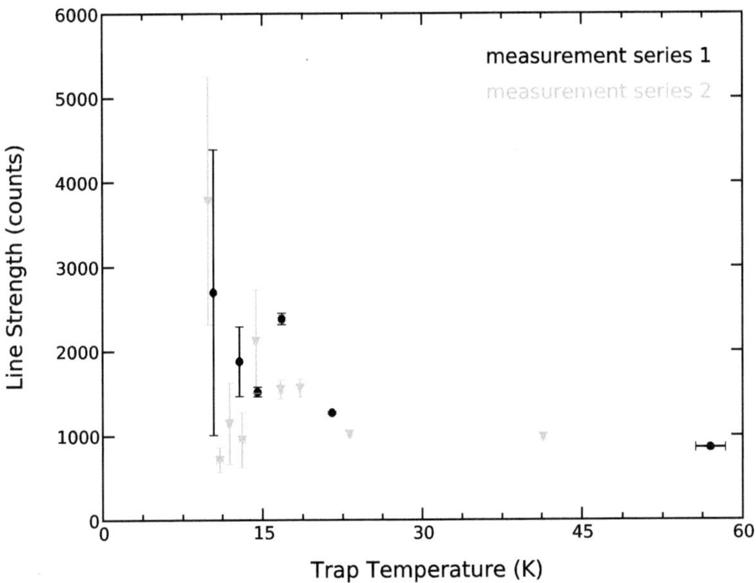

Figure 8.20: Dependence of the line strength N_0 on the trap temperature. Two measurement series at similar experimental conditions are shown.

Line Strength:

At 10 K, the investigated line at 3041.6355 cm^{-1} is one of the strongest (comparing the absolute height of the lines of different transitions), so its lower transition state is expected to be one of the lowest rotational states of CH_5^+. With increasing temperature its population will then be decreased, as higher energy levels get populated. The thermal population of the lower transition level might influence the line strength. The measurements shown in Figure 8.20 do not show any trend and the data taken at low temperatures have rather large errors. This is again due to the low saturations of many of these lines, leading to confusion between the fit parameters strength, saturation, and width, as can be seen from Figure 8.21.

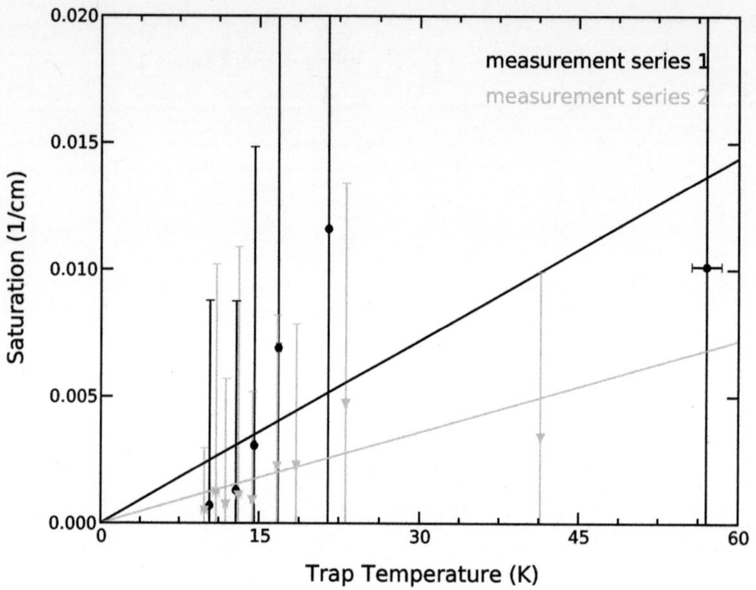

Figure 8.21: Dependence of the saturation a on the trap temperature. Two measurement series at similar experimental conditions are shown. The solid lines are not fits but serve as guide to the eye for the expected trend.

Saturation:

The change of level population with the temperature might influence the saturation of the lines. It was a main goal of the experiments on CH_5^+ to get a better understanding of this part of the process. In Figure 8.21 seems to be a clear trend of saturation increasing with temperature. However, the quality of the measurements is again too low to derive a quantitative description of the correlation saturation-temperature.

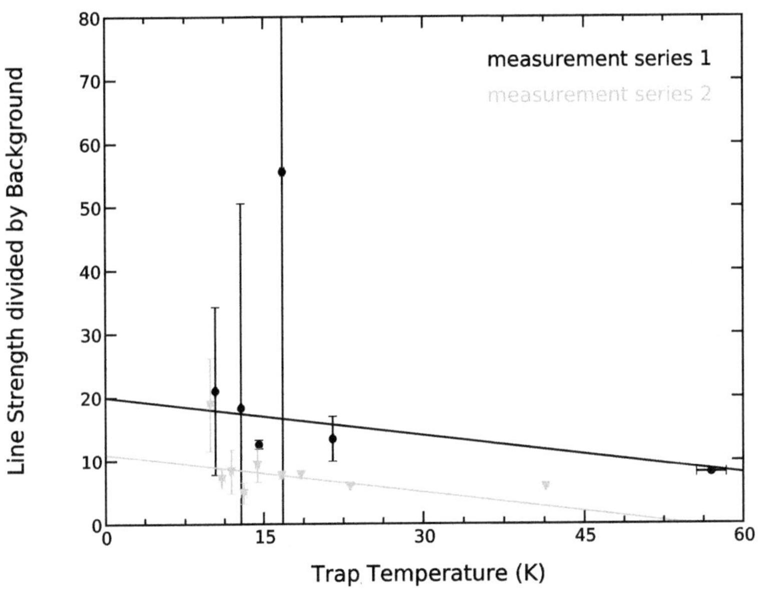

Figure 8.22: Dependence of the ratio line strength to background $\left(\frac{N_0}{C}\right)$ on the trap temperature. Two measurement series at similar experimental conditions are shown. The solid lines are not fits but serve as guide to the eye for the expected trend.

Line Strength to Background Ratio:

The trap temperature will most likely influence all fit parameters, because it changes the velocity distribution of the ions as well as the distribution over the internal energy levels. Therefore, at higher temperature, more ions will have enough energy to react without laser absorption, leading to a higher background. Because of this effect, the ratio $\frac{\text{line strength}}{\text{background}}$ might be less helpful than for the other experimental parameters. Figure 8.22 shows that this ratio decreases with increasing temperature. This effect might be due to an increase of the background with the temperature or to a decrease of the population of the pumped level.

8.4   Conclusion on Line Profiles

The lines observed with LIR spectroscopy often show deviations from a Gaussian line profile. Their profile can be described by a saturated line which results when a Gaussian line profile is inserted into Lambert-Beer's law. Such functions usually describe absorption processes, where the number of photons is limited in comparison to the number of absorbing molecules; they also work, when the number of absorbers is limited compared to the number of photons.

Despite the good qualitative understanding of the processes leading to the observed line profiles, there is still a lack of understanding when it comes to a quantitative description. Numerical simulations on a system similar to H_2D^+ and test measurements on CH_5^+ have been performed to improve the quantitative description of the LIR process.

There are strong hints both from the experiment and from the simulation, that the saturation of the lines becomes stronger with increasing time and laser power. It should depend in the same way on the Einstein B coefficient of the observed transition, but measurements on that effect were not possible, since the Einstein B coefficients for the transitions of CH_5^+ are not known yet. The saturation may also depend on the trap temperature, but experiments were not able to prove this.

The strength of the lines N_0 is most likely proportional to the number of ions inserted in the trap, but the proportionality factor is far below one (one would mean that all inserted parent ions can react to form the products). It also seems to depend on the investigated transition. For example for CH_5^+, the fits showed much lower strengths for the weaker line in the left slope of the transition that was of main interest to the experiments performed for this work (the line strength describes the number of ions available for the reaction, not the height of the observed line, and thus should be independent of the pumped transition). The line strength however seems to be independent from the other experimental parameters laser power, storage time and trap temperature.

As expected, the line width seems to depend only on the trap temperature. This allows to determine the translational temperature of the ions also from the width of observed lines, which do not show a Gaussian profile.

The measurements shown here were able to shed some light on the issue of understanding the line profiles, but it is recommended to repeat them with another reaction system. The spectroscopy of CH_5^+ is not yet understood, therefore it is not possible to calculate important information as the thermal population of the lower transition state at the experimental temperature or the Einstein B coefficient for the observed transition.

Additionally, the reaction gas CO_2 had to be inserted in the trap via a pulsed valve to reduce the amount of frozen gas on the surfaces inside the apparatus. Since the theory leading to the line profile given in Eq. (8.3) considers an amount

Table 8.1: Summary of the experiments on line profiles:
Left: Parameters that enter the fit function (8.3).
Top: Experimental parameters that most likely influence the observed line profile. The dependencies were derived from the graphs shown in Figures 8.7 to 8.22, the colours and symbols shown here refer to the coding of the figures.

	number of parent ions at zero time	laser power	time	trap temperature
$\dfrac{\text{line strength}}{\text{background}}$	independent	most likely independent	inconclusive	inconclusive
line strength	most likely proportional	most likely independent	independent	inconclusive
saturation $(a = At\sigma\sqrt{2\pi})$	independent	proportional	most likely proportional	maybe proportional
line width	most likely independent	most likely independent	independent	most likely $\propto \sqrt{T_{trap}}$

of reaction gas that does not change with time, this is likely to cause further complications.

The best possible test case for the investiagtion of saturation effects is H_2D^+, since its spectroscopy is well understood and the reaction with H_2 is comparatively easy to handle in LIRTrap experiments. Another improvement with respect to the experiments on CH_5^+ would be the fact, that for spectroscopy on H_2D^+ a diode laser is available (see Chapter 4.4). For this laser the frequency and the output power can be chosen reproducibly and are stable over the whole measurement series, which is not possible for the OPO. Unfortunately, it was not available for this work, as the switch to another reaction system and radiation source takes at least two weeks and would have significantly disturbed to ongoing measurement on CH_5^+.

When performing new test measurements towards an understanding of the line profiles of LIR lines, one should use preferably conditions, at which the lines are already saturated significantly (saturation $\frac{a}{\sigma} > 1$). This way, correlations between the fit parameters can be avoided. Since the saturation parameter a

makes lines higher and broader, it can not be fitted independently from these parameters for lines that are very close to a Gaussian profile.

The measurements and theoretical considerations presented in this chapter suggest that higher saturations can be reached by longer storage times, higher laser power and by optimizing the trap temperature (for the trap temperature, the saturation seems to have a maximum, probably at the temperature, where the thermal population of the pumped level is maximal). Longer storage times can be easily reached. For cryogenic temperatures, usually storage times up to several seconds are not a problem. However, they increase the time consumption of the measurements, which might cause other problems when experimental parameters show drifts. Higher laser power does not come with such disadvantages, but is not an option with the radiation sources available for this work, who were already operated at their output power limits for some of the measurements presented here. The investigated CH_5^+ line showed higher saturations for higher temperatures. Probably, there will be a different temperature for the maximal saturation for each line, given by the thermal population of the levels (compare Equation 8.17).

Chapter 9

Conclusion and Outlook

A variety of ion-neutral reaction systems has been investigated in this work with different aims. All of their equilibria depend very sensitively on the translational and internal energies of the reaction partners at cryogenic temperatures (down to 10 K). This sensitivity is a powerful tool, that can be applied to get plenty of information about the involved reaction partners and reaction mechanisms.

The most straightforward application is spectroscopy of ions via laser induced reactions. In this work, the LIR method was used to obtain high resolution infrared spectra of CH_2D^+. Those spectra enable high quality predictions of pure rotational transitions of the ion, which are the basis for its identification in the interstellar medium.

The lines obtained with LIR spectroscopy often show deviations from a Gaussian line profile. The profiles observed can be described by the combination of a Gaussian profile and Lambert-Beer's Law, so that precise determination of the center frequency is not hindered. However, the description of the observed lines is just phenomenological so far. The proper function is available and can be fitted to the observed lines, but the correlation between fit and experimental parameters is not understood, yet.

First steps towards a better understanding of the involved processes and towards their quantitative description have been made in this work. A rather simple numerical model was able to reproduce the observed line profiles as well as the expected time dependence of the height of the lines. Test measurements on the correlation of experimental and fit parameters were performed on the CH_5^+ ion. Some hints towards the correlations could be derived from the test measurements and from the numerical simulations. However, CH_5^+ is a rather unlucky choice for such tests, as its spectroscopy is not yet understood, the radiation source is quite reliable in frequency, but not in power, and the experimental procedures have to be modified for CH_5^+ in a way that prevents measurements on the temporal evolution of the LIR lines.

Future investigations should be performed on the H_2D^+ ion. Its rovibrational spectroscopy is well understood, the experiment is comparatively easy to handle and a radiation source is available, that is precisely and reproducibly tunable in frequency and output power.

The reaction system $N^+ + H_2 \rightleftharpoons NH^+ + H$ has been investigated in great detail. The reaction of N^+ with H_2 is the first step in the formation of interstellar ammonia. It has been investigated by several groups in the last three decades, but is still not completely understood. It is known to require energy, but whether this is due to endothermicity or to a barrier, is still an open question.

The rotational energy of ortho hydrogen is known to enhance the reaction at low temperatures. This effect will make the reaction with N^+ ions a very sensitive test mechanism for the ortho-to-para ratio in a given hydrogen sample, once it is well enough understood. So far, it is not known to what degree the energy of the fine-structure levels of N^+ is also able to support the reaction.

In the experiments presented in this work, many parameters were observed to influence the measured rate coefficients. The observed effects strengthen the theory that not only ortho and para hydrogen require state specific reaction rates, but also the fine-structure states of N^+. Some experimental results can be taken as a hint that the lowest fine-structure state is reacting much slower than the first excited one, and that the second excited fine-structure state is reacting very slowly or not at all. However, the dependence of the rate coefficient on the experimental parameters is not well enough understood to always get reproducible results. Some of the observed effects require a fine-structure population of N^+ that quickly thermalizes in the ion trap for fitting the data. Some effects can be explained only by models that include slow cooling of the fine-structure population to non-thermal distributions, which makes physically no sense. Thus, further work on this reaction system is required.

State specific rate coefficients are the aim not only for the reaction of N^+ with H_2, but for many ion-neutral reaction systems. Very promising is the system $H_2D^+ + H_2$. At ten Kelvin, most of the population will be in four states for the ions and two states for the neutrals. When the hydrogen is pure para hydrogen, the number of states is reduced to one for the neutrals. Application of double-resonance LIR spectroscopy (as described in [17]) will then give insight, how much each of the rotational levels of H_2D^+ contributes to the formation of LIR signal and background. From this, state-specific rate coefficients might be derived.

It is crucial to know the ortho fraction remaining in para hydrogen samples to investigate endothermic reactions at cryogenic temperatures. Therefore, part of this work was focused on the determination of the para hydrogen purity and on improvements on the para hydrogen generator.

Besides the reaction system $N^+ + H_2$, the system $H_2D^+ + H_2$ can serve as a test mechanism for the p-H_2 purity. It also comes with some difficulties, as the reaction equilibrium also depends on the HD amount in the reactant gas. First tests result in ortho fractions of less than 0.2 %, which is close to the desired purity of better than 99.9 % p-H_2.

Raman spectroscopy is another test method for the quality of p-H_2. It requires higher amounts of hydrogen than the original setup of the para generator could provide. Therefore, the apparatus was modified to produce up to 750 mbar of para hydrogen. These gas samples are stored in special bottles to prevent back conversion to normal hydrogen. Hence, now the same sample can be used in several ion trap experiments without the risk of fluctuations (in the H_2 flux or the converter temperature) during the production process.

Raman spectroscopy itself turned out to have a lower detection limit of 0.5 % o-H_2, which is not sufficient to prove the desired quality of the para hydrogen samples. The aim is to use Raman measurements after longer storage times. From the observation of the back conversion processes, not only the time constant for the conversion can be derived, but also the ortho fraction immediately after production of the sample. So far, the experimental drifts of the Raman spectrometer are too large, to apply this method with sufficient accuracy. More investigations are required to determine the origin of these drifts.

Over the last years, two new ion trap experiments have been built in the Cologne laboratories. The construction of the photomultiplier readout system for the Daly detectors of those experiments was also part of the work presented here. A housing for the electronics was developed, that prevents ambient light from entering the PMT and that contains all necessary electronics for power supply and readout in as little space as possible.

Bibliography

[1] Emmanuelle Lescop. *Laserinduzierte Prozesse im System* $C_2H_2^+ + H_2$. PhD thesis, Technische Universität Chemnitz, 2000.

[2] Stephan Schlemmer and Oskar Asvany. Laser Induced Reactions in a 22-Pole Ion Trap. In *J. Phys. Conf. Ser.*, volume 4, pages 134–141, 2005.

[3] N. R. Daly. Scintillation type mass spectrometer ion detector. *Rev. Sci. Instr.*, 31:264–267, 1960.

[4] Dieter Gerlich. Inhomogeneous RF fields: A versatile Tool for the Study of processes with Slow Ions. In Cheuk-Yiu Ng and Michael Baer, editors, *Adv. Chem. Phys.: State-Selected and State-to-State Ion-Molecule Reaction Dynamics*, volume LXXXII, pages 1–176. Wiley, New York, 1992.

[5] Annette Sorgenfrei. *Ion-Molekül-Reaktionen kleiner Kohlenwasserstoffe in einem gekühlten Ionen-Speicher*. PhD thesis, University of Freiburg, Germany, 1994.

[6] Oskar Asvany, Frank Bielau, Damian Moratschke, Jürgen Krause, and Stephan Schlemmer. New design of a cryogenic linear RF multipole trap. *Rev. Sci. Instr.*, 81:076102, 2010.

[7] Sabrina Gärtner. Charakterisierung einer Ionenwolke in einem kalten 22-Pol Ionenspeicher. Diplomarbeit, Universität zu Köln, 2009.

[8] O. Asvany, S. Schlemmer, and D. Gerlich. Deuteration of CH_n^+ (n=3-5) in collisions with HD measured in a low temperature ion trap. *Astrophys. J.*, 617:658–692, 2004.

[9] Edouard Hugo, Oskar Asvany, and Stephan Schlemmer. $H_3^+ + H_2$ isotopic system at low temperatures: Microcanonical model and experimental study. *J. Chem. Phys.*, 130:164302, 2009.

[10] Lars Kluge, Sabrina Gärtner, Sandra Brünken, Oskar Asvany, Dieter Gerlich, and Stephan Schlemmer. Transfer of a proton between H_2 and O_2. *Phil. Trans. R. Soc. A*, 370(1978):5041–5054, 2012.

[11] Oskar Asvany, Thomas Giesen, Britta Redlich, and Stephan Schlemmer. Experimental determination of the ν_5 cis-bending vibrational frequency in ground state $(X\,{}^2\Pi_u)\,C_2H_2^+$ using laser induced reactions. *Phys. Rev. Lett.*, 94:073001, 2005.

[12] Oskar Asvany, Padma Kumar P, Britta Redlich, Ilka Hegemann, Stephan Schlemmer, and Dominik Marx. Understanding the Infrared Spectrum of Bare CH_5^+. *Science*, 309:1219–1222, 2005.

[13] Sabrina Gärtner, Jürgen Krieg, Andre Klemann, Oskar Asvany, Sandra Brünken, and Stephan Schlemmer. High-Resolution Spectroscopy of CH_2D^+ in a Cold 22-Pole Trap. *J. Phys. Chem. A*, in press:10.1021/jp400258e.

[14] Sabrina Gärtner, Jürgen Krieg, Andre Klemann, Oskar Asvany, and Stephan Schlemmer. Rotational Transitions of CH_2D^+ determined by high-resolution ir spectroscopy. *Astron. Astrophys.*, 516:L3, 2010.

[15] Stephan Schlemmer, Thomas Kuhn, Emmanuelle Lescop, and Dieter Gerlich. Laser excited N_2^+ in a 22-pole ion trap: experimental studies of rotational relaxation processes. *Int. J. Mass Spectrom.*, 185:589–602, 1999.

[16] Stephan Schlemmer, Emmanuelle Lescop, Jan v. Richthofen, and Dieter Gerlich. Laser Induced Reactions in a 22-Pole Trap: $C_2H_2^+ + h\nu_3 + H_2 \rightarrow C_2H_3^+ + H$. *J. Chem. Phys.*, 117:2068–2075, 2002.

[17] Oskar Asvany, Oliver Ricken, Holger S. P. Müller, Martina C. Wiedner, Thomas Giesen, and Stephan Schlemmer. High-Resolution Rotational Spectroscopy in a cold Ion Trap: H_2D^+ and D_2H^+. *Phys. Rev. Lett.*, 100:233004, 2008.

[18] Oskar Asvany, Edouard Hugo, Frank Müller, Frank Kühnemann, Stephan Schiller, Jonathan Tennyson, and Stephan Schlemmer. Overtone spectroscopy of H_2D^+ and D_2H^+ using laser induced reactions. *J. Chem. Phys.*, 127:154317, 2007.

[19] Sven Fanghänel. Pulshöhenverteilungen eines empfindlichen Ionendetektors. Staatsexamensarbeit, Universität zu Köln, 2012.

[20] Juliane Gerke. Production and Characterization of Para Hydrogen. Masters thesis, University of Cologne, 2012.

[21] Dennis Bing. Kontrollierte Parawasserstofferzeugung zur Untersuchung der Spinsymmetrie bei gespeicherten Wasserstoffmolekülionen - Aufbau und Charakterisierung eines Parawasserstoff-Konverters. Diplomarbeit, Ruprecht-Karls-Universität Heidelberg, 2006.

[22] R.E. Honig. A fairy tale: Vapor pressure data of the elements.

[23] M. Cordonnier, D. Uy, R. M. Dickson, K. E. Kerr, Y. Zhang, and T. Oka. Selection rules for nuclear spin modifications in ion-neutral reactions involving H_3^+. *J. Chem. Phys.*, 113:3181–3193, August 2000.

[24] F. Grussie, M. H. Berg, K. N. Crabtree, S. Gärtner, B. J. McCall, S. Schlemmer, A. Wolf, and H. Kreckel. The Low-temperature Nuclear Spin Equilibrium of H_3^+ in Collisions with H_2. *Astrophys. J.*, 759:21, 2012.

[25] Dieter Gerlich, Eric Herbst, and Evelyne Roueff. H_3^+ + HD → H_2D^+ + H_2: low-temperature laboratory measurements and interstellar implications. *Planetary and Space Science*, 50:1275–1285, 2002.

[26] Edouard Hugo. *The H_3^+ + H_2 isotopic system*. PhD thesis, University of Cologne, 2009.

[27] N. G. Adams and D. Smith. A Laboratory Study of the Reaction H_2D^+ + H_2 ↔ H_3^+ + HD: The Electron Densities and the Temperratures in Interstellar Clouds. *Astrophys. J.*, 248:373–379, 1981.

[28] Kevin Giles, Nigel G. Adam, and David Smiths. A Study of the Reactions of H_3^+, H_2D^+, HD_2^+, and D_3^+ with H_2, HD and D_2 Using a Variable-Temperature Selected Ion Flow Tube. *J. Phys. Chem.*, 96:7645–7650, 1992.

[29] T. Amano and J. K. G. Watson. Observation of the ν_1 fundamental band of H_2D^+. *J. Chem. Phys.*, 81:2869–2871, 1984.

[30] T. Amano. Difference-frequency laser spectroscopy of molecular ions with a hollow-cathode cell: extended analysis of the ν_1 band of H_2D^+. *J. Opt. Soc. Am. B*, 2:790–793, 1985.

[31] T. Amano and T. Hirao. Accurate rest frequencies of submillimeter-wave lines of H_2D^+ and D_2H^+. *J. Mol. Spectrosc.*, 233:7–14, 2005.

[32] T. Amano. Submillimetre-wave lines of H_2D^+ and D_2H^+ as probes into chemistry in cold dark clouds. *Phil. Trans. R. Soc. A*, 364:2943–2952, 2006.

[33] T. Amano. Submillimetre-wave spectrum of CH_2D^+. *Astron. Astrophys.*, 516:L4, 2010.

[34] Oskar Asvany, Edouard Hugo, Serjoscha Wahed, and Stephan Schlemmer. Probing the low-temperature population of H_2D^+. In *J. Phys. Conf. Ser.*, volume 192, page 012010, 2009.

[35] M. Bogey, C. Demuynck, M. Denis, J. L. Destombes, and B. Lemoine. Laboratory measurement of the 1_{01}– 1_{11} submillimeter line of H_2D^+. *Astron. Astrophys.*, 137:L15–L16, 1984.

[36] Michal Fárník, Scott Davis, Maxim A. Kostin, Oleg L. Polyansky, Jonathan Tennyson, and David Nesbitt. Beyond the Born-Oppenheimer approximation: High-resolution overtone spectroscopy of H_2D^+ and D_2H^+. *J. Chem. Phys.*, 116:6146–6158, 2002.

[37] S. C. Foster, A. R. W. McKellar, I. R. Peterkin, J. K. G. Watson, F. S. Pan, M. W. Crofton, R. S. Altman, and T. Oka. Observation and analysis of the ν_2 and ν_3 fundamental bands of the H_2D^+ ion. *J. Chem. Phys.*, 84:91–99, 1986.

[38] I. N. Kozin, O. L. Polyansky, and N. F. Zobov. Improved Analysis of the Experimental Data on the H_2D^+ and D_2H^+ Absorption Spectra. *J. Mol. Spectrosc.*, 128:126–134, 1988.

[39] Oleg L. Polyansky and Jonathan Tennyson. *Ab initio* calculation of the rotation-vibration energy levels of H_3^+ and its isotopomers to spectroscopic accuracy. *J. Chem. Phys.*, 110:5056–5064, 1999.

[40] Stephan Schlemmer, Oskar Asvany, Edouard Hugo, and Dieter Gerlich. Deuterium Fractionation and Ion-Molecule Reactions at low Temperatures. In D.C. Lis, G.A. Blake, and E. Herbst, editors, *IAU Symposium 231*, 2006.

[41] J.-T. Shy, John W. Farley, and William H. Wing. Observation of the infrared spectrum of the triatomic molecular ion H_2D^+. *Phys. Rev. A*, 24:1146–1149, 1981.

[42] Hugh E. Warner, William T. Conner, Rudolph H. Petrmichl, and R. Claude Woods. Laboratory detection of the $1_{01} \leftarrow 1_{11}$ submillimeter wave transition of the H_2D^+ ion. *J. Chem. Phys.*, 81:2514, 1984.

[43] R. T. Boreiko and A. L. Betz. A Search for the rotational transitions of H_2D^+ at 1370 GHz and H_3O^+ at 985 GHz. *Astrophys. J.*, 405:L39–L42, 1993.

[44] P. Caselli, F. F. S. van der Tak, C. Ceccarelli, and A. Bacmann. Abundant H_2D^+ in the pre-stellar core L1544. *Astron. Astrophys.*, 403:L37–L41, 2003.

[45] C. Ceccarelli, C. Dominik, B. Lefloch, P. Caselli, and E. Caux. Detection of H_2D^+: measuring the midplane degree of ionization in the disks of DM Tauri and TW Hydrae. *Astrophys. J.*, 607:L51–L54, 2004.

[46] J. Cernicharo, E. Polehampton, and J. R. Goicoechea. Far-Infrared detection of H_2D^+ toward Sgr B2. *Astrophys. J*, 657:L21–L24, 2007.

[47] Ronald Stark, Floris F.S. van der Tak, and Ewine F. van Dishoeck. Detection of interstellar H_2D^+ emission. *Astrophys. J.*, 521:L67–L70, 1999.

[48] P. F. Bernath. *Spectra of Atoms and Molecules.* Oxford University Press, 1995.

[49] Python, http://www.python.org/.

[50] I. Zymak, M. Hejduk, D. Mulin, R. Plašil, J. Glosík, and D. Gerlich. Low-Temperature Ion Trap Studies of $N^+(^3P_{ja}) + H_2(j) \rightarrow NH^+ + H$. *Astrophys. J.*, 768:86, 2013.

[51] Dieter Gerlich. Experimental Investigation of Ion-Molecule Reactions Relevant to Interstellar Chemistry. *J. Chem. Soc. Faraday Trans.*, 89:2199–2208, 1993.

[52] T. Mayer-Kuckuk. *Atomphysik.* Teubner Studienbücher, 1985.

[53] Dieter Gerlich. Private Communication, 2013.

[54] data taken from http://webbook.nist.gov/chemistry.

[55] N. Wiberg A. F. Holleman, E. Wiberg. *Lehrbuch der Anorganischen Chemie.* Walter de Gruyter, 1995.

[56] M. Rösslein, M. F. Jagod, C. M. Gabrys, and T. Oka. Laboratory Infrared Spectra of CH_2D^+ and $HCCD^+$ and Predicted Microwave Transitions. *Astrophys. J.*, 382:L51–L53, 1991.

[57] Mary-Frances Jagod, Matthias Rösslein, Charles M. Gabrys, and Takeshi Oka. Infrared Spectroscopy of Carbo-ions: The ν_1 and ν_4 Bands of CH_2D^+, and the ν_1 Band of CHD_2^+. *J. Molec. Spectroscopy*, 153:666–679, 1992.

[58] E. Roueff, M. Gerin, D. C. Lis, A. Wootten, N. Marcelino, J. Cernicharo, and B. Tercero. CH_2D^+, the Search for the Holy Grail. *J. Phys. Chem. A*, in press:10.1021/jp400119a.

[59] PGOPHER, a Program for Simulating Rotational Structure, C. M. Western, University of Bristol, http://pgopher.chm.bris.ac.uk.

[60] Jürgen Krieg. *High-Resolution Infrared Spectroscopy of Transient Molecules - Development of Broadband Optical Parametric Oscillators.* PhD thesis, Universität zu Köln, 2011.

[61] Jürgen Krieg, Andre Klemann, Imke Gottbehüt, Sven Thorwirth, T.F. Giesen, and Stephan Schlemmer. A continuous-wave optical parametric oscillator around $5\mu m$ wavelength for high-resolution spectroscopy. *Rev. Sci. Instr.*, 82:063105, 2011.

[62] S. Saupe, M. H. Wappelhorst, B. Meyer, W. Urban, and A. G. Maki. Sub-Doppler Heterodyne Frequency Measurements near 5 μm with a CO-Laser Sideband System: Improved Calibration Tables for Carbonyl Sulfide Transitions. *J. Molec. Spectr.*, 175:190–197, 1996.

[63] G. Guelachvili and K. N. Rao. *Handbook of Infrared Standards II.* Academic Press, Inc., San Diego, 1993.

[64] R. A. Toth. $2\nu_2 - \nu_2$ and $2\nu_2$ bands of $H_2^{16}O$, $H_2^{17}O$ and $H_2^{18}O$: line positions and strenghts. *J. Opt. Soc. Am. B*, 10:1526–1544, 1993.

[65] H. M. Pickett. The Fitting and Prediction of Vibration-Rotation Spectra with Spin Interactions. *J. Molec. Spectroscopy*, 148:371–377, 1991.

[66] P. Langevin. Une formule fondamentale de théorie cinétique. *Annales de Chimie et de Physique*, 5:245, 1905.

[67] Oskar Asvany, Jürgen Krieg, and Stephan Schlemmer. Frequency comb assisted mid-infrared spectroscopy of cold molecular ions. *Rev. Sci. Instr.*, 83:093110, 2012.

Appendix A

A.1   Python Simulation for LIR Line Profiles

A.1.1   Main Program

```python
# load required packages
import math
import matplotlib.pyplot as plt
import os
import sys

# import my own functions
sys.path.append("functions")        # set the directory containing the functions
from function_frequency_dependence import frequency_dependence
from function_time_dependence import time_dependence
from function_write_to_file import write_to_file
from function_plot_simulation import plot_simulation

##########################################

# plot settings
fig = plt.figure(figsize=plt.figaspect(2.))          # ratio of height to width
fig.subplots_adjust(wspace=0.5, hspace=0.5)       # have more space between
subplots
plt.rc("xtick", labelsize = 3)        # fontsize for tics
plt.rc("ytick", labelsize = 3)

##########################################

# simulation settings
delt=0.0001        # time steps in sec :  0.1 ms
steps=35000        # total number of time steps
printsteps=10       # number of time points to be written into file
points=10        # number of frequency points on each side of maximum

# experimental conditions
```

```python
T_ion=30         # corrected ion temperature in K
T_trap=20         # corrected gas temperature in K
p_trap=2.76e-8        # H2-pressure in mbar
p_laser=6.0*1e-3        # laser power in W
Ions=1000         # total number of H2D+ ions at zero time

# probability of decay to lower level instead of reaction during collision
decay = 0.8

# go to the directory in which to write the output
if "results_2" in os.listdir(".") :          # does directory "results_2" exist
in current directory?
   pass        # if so, continue
else:
   os.mkdir("results_2")          # if not, make it
   os.chdir("results_2")          # go into directory "results_2"

if "decay_%.1f_-_T_ion_%dK_-_T_trap_%dK_-_p_trap_%.2embar_-_p_laser
_%.1fmW_-_Ions_%d" %(decay, T_ion, T_trap, p_trap, p_laser*1e3, Ions)
in os.listdir(".") :
   pass
else:
   os.mkdir("decay_%.1f_-_T_ion_%dK_-_T_trap_%dK_-_p_trap_%.2embar_-_p_laser
_%.1fmW_-_Ions_%d" %(decay, T_ion, T_trap, p_trap, p_laser*1e3, Ions))
   os.chdir("decay_%.1f_-_T_ion_%dK_-_T_trap_%dK_-_p_trap_%.2embar_-_p_laser
_%.1fmW_-_Ions_%d" %(decay, T_ion, T_trap, p_trap, p_laser*1e3, Ions))

# constants
Pi=3.1415927
h=6.6261e-34        # in Js
c=3e8        # in m/s
Area=0.5e-4        # cross-sectional area of ion cloud and laser-beam in m^2
f_cal=290        # calibration factor for the pressure in the trap
u=1.661e-27        # atomic mass unit in kg
k_B=1.381e-23        # Boltzmann constant in J/K

for l in range(4):
   L=l+1
   print ("\n---------------------\nLevel %d:" %L)
# File to write results in
   filename = "level_%d.dat" %L

#########################################
# constants for the transition used
   if L == 1:
      Lambda=1546.4242*1e-9        # wavelength in m
      A_ul=4.10        # Einstein-A-coefficient
```

```python
        gl=1          # degeneracy of lower rot level
    if L == 2:
        Lambda=1540.5118*1e-9        # wavelength in m
        A_ul=4.49          # Einstein-A-coefficient
        gl=5          # degeneracy of lower rot level
    if L == 3:
        Lambda=1548.2187*1e-9        # wavelength in m
        A_ul=2.47          # Einstein-A-coefficient
        gl=5          # degeneracy of lower rot level
    if L == 4:
        Lambda=1569.9918*1e-9        # wavelength in m
        A_ul=6.04          # Einstein-A-coefficient
        gl=3          # degeneracy of lower rot level
###########################################

    m=4*u          # mass of an H2D+ ion
    nu=c/Lambda          # frequency in Hz
    B_emis=(gl*A_ul*c**3)/(8*Pi*h*nu**3)          # Einstein-B-coefficient for
emission (degeneracy of rot-levels)
    B_abs=B_emis*3/gl          # Einstein-B-coefficient for absorption (degeneracy
of rot-levels)
    w=((k_B*T_ion)/(m*c**2))**0.5*Lambda          # doppler width of transition
    w_nu=((k_B*T_ion)/(m*c**2))**0.5*nu          # doppler width of transition

# energy levels (in Kelvin)
    E_L1=0          # lower level 1
    E_L2=65.77          # lower level 2
    E_L3=86.38          # lower level 3
    E_L4=104.26          # lower level 4
    E_R=171.          # average for other levels (J=2)
    E_U=h*nu/k_B          # upper level
    print ("E_U = %.2f" %(E_U))

# H2 number-density
    n_H2=4.18e17*f_cal*p_trap/T_trap**0.5
    print (n_H2)

# ion types
    N_G=[]
    N_P=[]
    N_L2=[]
    N_L3=[]
    N_L4=[]
    N_R=[]
    N_U=[]
    N_L1=[]
```

```python
# reacting ions
   del_Pu = []        # H2D+ + h*nu -> H2D+*
   del_21 = []        # 2 -> 1
   del_23 = []        # 2 -> 3
   del_12 = []        # 1 -> 2
   del_32 = []        # 3 -> 2
   del_34 = []        # 3 -> 4
   del_43 = []        # 4 -> 3
   del_4R = []        # level 4 to remaining levels
   del_f = []       # H2D+* + H2 -> H3+ + HD
   del_Pd = []        # H2D+* + h*nu -> H2D+
   del_UR = []        # upper level to remaining levels
   del_R4 = []        # remaining level to level 4
   del_RU = []        # remaining levels to upper level

# ratecoefficients in cm^3/s
   k_f=1.79e-9        # ratecoefficient for reaction from upper level to product
(Langevin)

# rates in 1/s
   K_f=k_f*n_H2*(1-decay)        # reactionrate from upper level to product
   K_Pu=(B_abs*p_laser)/(w_nu*Area*c)        # H2D+ + h*nu -> H2D+*
   K_Pd=(B_emis*p_laser)/(w_nu*Area*c)        # H2D+* + h*nu -> H2D+
   K_UR=k_f*n_H2*decay        # upper level to remaining levels
   K_RU=K_UR*math.exp((E_R-E_U)/T_ion)        # remaining levels to upper level
   K_R4=k_f*n_H2*decay        # remaining levels to lower level 4
   K_4R=K_R4*math.exp((E_L4-E_R)/T_ion)    .    # lower level 4 to remaining
levels
   K_43=k_f*n_H2*decay        # lower level 4 to lower level 3
   K_34=K_43*math.exp((E_L3-E_L4)/T_ion)        # lower level 3 to lower level
4
   K_32=k_f*n_H2*decay        # lower level 3 to lower level 2
   K_23=K_32*math.exp((E_L2-E_L3)/T_ion)        # lower level 2 to lower level
3
   K_21=k_f*n_H2*decay        # lower level 2 to lower level 1
   K_12=K_21*math.exp((E_L1-E_L2)/T_ion)        # lower level 1 to lower level
2

################   simulation   ###############


#count entries generated by for-loops
   i_liste = []
   j_liste = []

# time
   time = []
```

```python
# frequency dependence
    j = 0          # counter
    y2 = []
    gauss = []

    print ("\nFrequency Iteration:")

    for f in range (-points, points+1) :
        frequency_dependence(Ions, time, j_liste, j, gauss, y2, w, f, Pi, del_Pu,
del_21, del_23, del_12, del_32, del_34, del_43, del_4R, del_f, del_Pd, del_UR,
del_R4, del_RU, N_P, N_U, N_R, N_L4, N_L3, N_L2, N_L1, N_G, E_L1, E_L2, E_L3,
E_L4, E_U, E_R, T_ion)

# time dependence
        for i in range(0,steps+1) :
            time_dependence(L, i, i_liste, j, steps, delt, time, gauss, del_Pu,
del_21, del_23, del_12, del_32, del_34, del_43, del_4R, del_f, del_Pd, del_UR,
del_R4, del_RU, N_P, N_U, N_R, N_L4, N_L3, N_L2, N_L1, N_G, K_f, K_Pu, K_Pd,
K_UR, K_RU, K_R4, K_4R, K_43, K_34, K_32, K_23, K_21, K_12)

        j += 1
        i_liste.append(j*steps+2)

#############################################

    print ("\nRates:\n K_f = %.2E\n K_Pu = %.2E\n K_Pd = %.2E\n K_UR = %.2E\n
K_RU = %.2E\n K_R4 = %.2E\n K_4R = %.2E\n K_43 = %.2E\n K_34 = %.2E\n K_32
= %.2E\n K_23 = %.2E\n K_21 = %.2E\n K_12 = %.2E\n" %(K_f, K_Pu, K_Pd, K_UR,
K_RU, K_R4, K_4R, K_43, K_34, K_32, K_23, K_21, K_12))

    print ("Ions at t=0:\n N_G = %d\n N_P = %d\n N_L2 = %d\n N_L3 = %d\n N_L4
= %d\n N_R = %d\n N_U = %d\n N_L1 = %d\n " %(N_G[0], N_P[0], N_L2[0], N_L3[0],
N_L4[0], N_R[0], N_U[0], N_L1[0]))

    print ("Transition Constants:\n Lambda = %.4E\n nu=c/Lambda = %.4E\n
B_emis=(gl*A_ul*c**3)/(8*Pi*h*nu**3) = %.4E\n B_abs=B_emis*3/gl = %.4E\n
w=((k_B*T_ion)/(m*c**2))**0.5*Lambda = %.4E\n w_nu=((k_B*T_ion)/(m*c**2))**0.5*nu
= %.4E \n" %(Lambda, nu, B_emis, B_abs, w, w_nu))

############### write data into file #############

    write_to_file(filename, time, j_liste, i_liste, N_P, N_L1, N_L2, N_L3,
N_L4, N_R, N_U, N_G, y2, gauss, Lambda, w, del_Pu, del_21, del_23, del_12,
del_32, del_34, del_43, del_4R, del_f, del_Pd, del_UR, del_R4, del_RU, j,
i, steps, printsteps)

################### plot ##################

    plot_simulation(fig, printsteps, points, w, K_Pu, decay, N_G, N_L1, N_L2,
```

```python
                                      N_L3, N_L4, L)

# save figure
fig.savefig("1_simulation_with_plot.pdf", format="pdf")

# show plots
plt.show()
```

A.1.2 Frequency Dependence

```python
import math

def frequency_dependence(Ions, time, j_liste, j, gauss, y2, w, f, Pi, del_Pu,
del_21, del_23, del_12, del_32, del_34, del_43, del_4R, del_f, del_Pd, del_UR,
del_R4, del_RU, N_P, N_U, N_R, N_L4, N_L3, N_L2, N_L1, N_G, E_L1, E_L2, E_L3,
E_L4, E_U, E_R, T_ion):

    ###########################################

    # zero time
    time.append(0)

# number of ions at zero-time
    # total number of ions (sum over all levels)
    N_G.append(Ions)
    # product ions
    N_P.append(0)
    # ions in level L2
    N_L2.append(N_G[0]/(math.exp(-E_L1/T_ion) + math.exp(-E_L2/T_ion)
+ math.exp(-E_L3/T_ion) + math.exp(-E_L4/T_ion) + math.exp(-E_R/T_ion)
+ math.exp(-E_U/T_ion))*math.exp(-E_L2/T_ion))
    # ions in level L3
    N_L3.append(N_G[0]/(math.exp(-E_L1/T_ion) + math.exp(-E_L2/T_ion)
+ math.exp(-E_L3/T_ion) + math.exp(-E_L4/T_ion) + math.exp(-E_R/T_ion)
+ math.exp(-E_U/T_ion))*math.exp(-E_L3/T_ion))
    # ions in level L4
    N_L4.append(N_G[0]/(math.exp(-E_L1/T_ion) + math.exp(-E_L2/T_ion)
+ math.exp(-E_L3/T_ion) + math.exp(-E_L4/T_ion) + math.exp(-E_R/T_ion)
+ math.exp(-E_U/T_ion))*math.exp(-E_L4/T_ion))
    # ions in remaining levels
    N_R.append(N_G[0]/(math.exp(-E_L1/T_ion) + math.exp(-E_L2/T_ion)
+ math.exp(-E_L3/T_ion) + math.exp(-E_L4/T_ion) + math.exp(-E_R/T_ion)
+ math.exp(-E_U/T_ion))*math.exp(-E_R/T_ion))
    # ions in upper level
    N_U.append(N_G[0]/(math.exp(-E_L1/T_ion) + math.exp(-E_L2/T_ion)
+ math.exp(-E_L3/T_ion) + math.exp(-E_L4/T_ion) + math.exp(-E_R/T_ion)
+ math.exp(-E_U/T_ion))*math.exp(-E_U/T_ion))
    # ions in level L1
    N_L1.append(N_G[0]-N_L2[0]-N_L3[0]-N_L4[0]-N_R[0]-N_U[0])

# reacting ions at zero-time
    del_Pu.append(0)
    del_21.append(0)
    del_23.append(0)
    del_12.append(0)
```

```python
        del_32.append(0)
        del_34.append(0)
        del_43.append(0)
        del_4R.append(0)
        del_f.append(0)
        del_Pd.append(0)
        del_UR.append(0)
        del_R4.append(0)
        del_RU.append(0)

    # frequency offset relative to center frequency
        y2.append(w/2*f)

    # normalised Gaussian at position y2
        gauss.append(math.exp(-y2[j]**2/(2*w**2))/(2*math.sqrt(2*Pi)))

        j_liste.append(j)
        print (" %d" %j)
```

A.1.3   Time Dependence

```python
def time_dependence(L, i, i_liste, j, steps, delt, time, gauss, del_Pu, del_21,
del_23, del_12, del_32, del_34, del_43, del_4R, del_f, del_Pd, del_UR, del_R4,
del_RU, N_P, N_U, N_R, N_L4, N_L3, N_L2, N_L1, N_G, K_f, K_Pu, K_Pd, K_UR,
K_RU, K_R4, K_4R, K_43, K_34, K_32, K_23, K_21, K_12):

    ##########################################

# pumping of ions from lower transition level to upper level by laser
    if L == 1:
        del_Pu.append(N_L1[int(j*(steps+2) + i)] * gauss[j] * K_Pu * delt)
    if L == 2:
        del_Pu.append(N_L2[int(j*(steps+2) + i)] * gauss[j] * K_Pu * delt)
    if L == 3:
        del_Pu.append(N_L3[int(j*(steps+2) + i)] * gauss[j] * K_Pu * delt)
    if L == 4:
        del_Pu.append(N_L4[int(j*(steps+2) + i)] * gauss[j] * K_Pu * delt)
# from level L1 to level L2
    del_12.append(N_L1[int(j*(steps+2) + i)] * K_12 * delt)
# from level L2
    # to level L1
    del_21.append(N_L2[int(j*(steps+2) + i)] * K_21 * delt)
    # to level L3
    del_23.append(N_L2[int(j*(steps+2) + i)] * K_23 * delt)
# from level L3
    # to level L2
    del_32.append(N_L3[int(j*(steps+2) + i)] * K_32 * delt))
    # to level L4
    del_34.append(N_L3[int(j*(steps+2) + i)] * K_34 * delt))
# from level L4
    # to level L3
    del_43.append(N_L4[int(j*(steps+2) + i)] * K_43 * delt))
    # to remaining levels
    del_4R.append(N_L4[int(j*(steps+2) + i)] * K_4R * delt))
# from upper level
    # reaction to product
    del_f.append(N_U[int(j*(steps+2) + i)] * K_f * delt))
    # pumping to lower transiton level by laser
    del_Pd.append(N_U[int(j*(steps+2) + i)] * gauss[j] * K_Pd * delt))
    # to remaining levels
    del_UR.append(N_U[int(j*(steps+2) + i)] * K_UR * delt))
# from remaining levels
    # to level L4
    del_R4.append(N_R[int(j*(steps+2) + i)] * K_R4 * delt))
    # to upper level
```

```python
    del_RU.append(N_R[int(j*(steps+2) + i)] * K_RU * delt))

###########################################

# results
    N_P.append(N_P[int(j*(steps+2) + i)] + del_f[int(j*(steps+2) + i)])
    N_U.append(N_U[int(j*(steps+2) + i)] + del_Pu[int(j*(steps+2) + i)]
- del_Pd[int(j*(steps+2) + i)] - del_f[int(j*(steps+2) + i)]
- del_UR[int(j*(steps+2) + i)] + del_RU[int(j*(steps+2) + i)])
    N_R.append(N_R[int(j*(steps+2) + i)] - del_RU[int(j*(steps+2) + i)]
+ del_UR[int(j*(steps+2) + i)]+ del_4R[int(j*(steps+2) + i)]
- del_R4[int(j*(steps+2) + i)])
    N_G.append(N_L1[int(j*(steps+2) + i)] + N_L2[int(j*(steps+2) + i)]
+ N_L3[int(j*(steps+2) + i)] + N_L4[int(j*(steps+2) + i)]
+ N_R[int(j*(steps+2) + i)] + N_U[int(j*(steps+2) + i)]
+ N_P[int(j*(steps+2) + i)])

    if L == 1:
        N_L4.append(N_L4[int(j*(steps+2) + i)]
- del_4R[int(j*(steps+2) + i)] + del_R4[int(j*(steps+2) + i)]
+ del_34[int(j*(steps+2) + i)] - del_43[int(j*(steps+2) + i)])
        N_L3.append(N_L3[int(j*(steps+2) + i)]
- del_34[int(j*(steps+2) + i)] + del_43[int(j*(steps+2) + i)]
+ del_23[int(j*(steps+2) + i)] - del_32[int(j*(steps+2) + i)])
        N_L2.append(N_L2[int(j*(steps+2) + i)]
- del_23[int(j*(steps+2) + i)] + del_32[int(j*(steps+2) + i)]
+ del_12[int(j*(steps+2) + i)] - del_21[int(j*(steps+2) + i)]
- del_Pu[int(j*(steps+2) + i)])
        N_L1.append(N_L1[int(j*(steps+2) + i)]
- del_12[int(j*(steps+2) + i)] + del_21[int(j*(steps+2) + i)]
+ del_Pd[int(j*(steps+2) + i)])

    if L == 2:
        N_L4.append(N_L4[int(j*(steps+2) + i)]
- del_4R[int(j*(steps+2) + i)] + del_R4[int(j*(steps+2) + i)]
+ del_34[int(j*(steps+2) + i)] - del_43[int(j*(steps+2) + i)])
        N_L3.append(N_L3[int(j*(steps+2) + i)]
- del_34[int(j*(steps+2) + i)] + del_43[int(j*(steps+2) + i)]
+ del_23[int(j*(steps+2) + i)] - del_32[int(j*(steps+2) + i)])
        N_L2.append(N_L2[int(j*(steps+2) + i)]
- del_23[int(j*(steps+2) + i)] + del_32[int(j*(steps+2) + i)]
+ del_12[int(j*(steps+2) + i)] - del_21[int(j*(steps+2) + i)]
- del_Pu[int(j*(steps+2) + i)] + del_Pd[int(j*(steps+2) + i)])
        N_L1.append(N_L1[int(j*(steps+2) + i)]
- del_12[int(j*(steps+2) + i)] + del_21[int(j*(steps+2) + i)])
```

```
    if L == 3:
        N_L4.append(N_L4[int(j*(steps+2) + i)]
- del_4R[int(j*(steps+2) + i)] + del_R4[int(j*(steps+2) + i)]
+ del_34[int(j*(steps+2) + i)] - del_43[int(j*(steps+2) + i)])
        N_L3.append(N_L3[int(j*(steps+2) + i)]
- del_34[int(j*(steps+2) + i)] + del_43[int(j*(steps+2) + i)]
+ del_23[int(j*(steps+2) + i)] - del_32[int(j*(steps+2) + i)]
+ del_Pd[int(j*(steps+2) + i)])
        N_L2.append(N_L2[int(j*(steps+2) + i)]
- del_23[int(j*(steps+2) + i)] + del_32[int(j*(steps+2) + i)]
+ del_12[int(j*(steps+2) + i)] - del_21[int(j*(steps+2) + i)]
- del_Pu[int(j*(steps+2) + i)])
        N_L1.append(N_L1[int(j*(steps+2) + i)]
- del_12[int(j*(steps+2) + i)] + del_21[int(j*(steps+2) + i)])

    if L == 4:
        N_L4.append(N_L4[int(j*(steps+2) + i)]
- del_4R[int(j*(steps+2) + i)] + del_R4[int(j*(steps+2) + i)]
+ del_34[int(j*(steps+2) + i)] - del_43[int(j*(steps+2) + i)]
+ del_Pd[int(j*(steps+2) + i)])
        N_L3.append(N_L3[int(j*(steps+2) + i)]
- del_34[int(j*(steps+2) + i)] + del_43[int(j*(steps+2) + i)]
+ del_23[int(j*(steps+2) + i)] - del_32[int(j*(steps+2) + i)])
        N_L2.append(N_L2[int(j*(steps+2) + i)]
- del_23[int(j*(steps+2) + i)] + del_32[int(j*(steps+2) + i)]
+ del_12[int(j*(steps+2) + i)] - del_21[int(j*(steps+2) + i)]
- del_Pu[int(j*(steps+2) + i)])
        N_L1.append(N_L1[int(j*(steps+2) + i)]
- del_12[int(j*(steps+2) + i)] + del_21[int(j*(steps+2) + i)])

    i_liste.append(i)
    time.append((i+1)*delt)
```

A.1.4   Write Results to File

```python
def write_to_file(filename, time, j_liste, i_liste, N_P, N_L1, N_L2, N_L3,
N_L4, N_R, N_U, N_G, y2, gauss, Lambda, w, del_Pu, del_21, del_23, del_12,
del_32, del_34, del_43, del_4R, del_f, del_Pd, del_UR, del_R4, del_RU, j,
i, steps, printsteps):

#########################################

# output
    output = []

# headline
    output.append("t\t")
    output.append("j\t")
    output.append("i\t")
    output.append("N_P\t")
    output.append("N_L1\t")
    output.append("N_L2\t")
    output.append("N_L3\t")
    output.append("N_L4\t")
    output.append("N_R\t")
    output.append("N_U\t")
    output.append("N_G\t")
    output.append("y2\t")
    output.append("gauss[y2]\t")
    output.append("Lambda\t")
    output.append("w\t")
    output.append("del_Pu\t")
    output.append("del_21\t")
    output.append("del_23\t")
    output.append("del_12\t")
    output.append("del_32\t")
    output.append("del_34\t")
    output.append("del_43\t")
    output.append("del_4R\t")
    output.append("del_f\t")
    output.append("del_Pd\t")
    output.append("del_UR\t")
    output.append("del_R4\t")
    output.append("del_RU\t")
    output.append("\n")

# data
    for j in j_liste :
        for i in range(printsteps+1) :
            output.append("%.4f\t" %(time[int(j*(steps+2) + i*steps/printsteps)]))
```

```python
        output.append("%d\t" %(j_liste[j]))
        output.append("%d\t" %(i_liste[int(j*(steps+2) + i*steps/printsteps)]))
        output.append("%.2E\t" %(N_P[int(j*(steps+2) + i*steps/printsteps)]))
        output.append("%.2E\t" %(N_L1[int(j*(steps+2) + i*steps/printsteps)]))
        output.append("%.2E\t" %(N_L2[int(j*(steps+2) + i*steps/printsteps)]))
        output.append("%.2E\t" %(N_L3[int(j*(steps+2) + i*steps/printsteps)]))
        output.append("%.2E\t" %(N_L4[int(j*(steps+2) + i*steps/printsteps)]))
        output.append("%.2E\t" %(N_R[int(j*(steps+2) + i*steps/printsteps)]))
        output.append("%.2E\t" %(N_U[int(j*(steps+2) + i*steps/printsteps)]))
        output.append("%.0f\t" %(N_G[int(j*(steps+2) + i*steps/printsteps)]))
        output.append("%.4E\t" %(y2[j]))
        output.append("%.2E\t" %(gauss[j]))
        output.append("%.7E\t" %(Lambda))
        output.append("%.4E\t" %(w))
        output.append("%.2E\t" %(del_Pu[int(j*(steps+2) + i*steps/printsteps)]))
        output.append("%.2E\t" %(del_21[int(j*(steps+2) + i*steps/printsteps)]))
        output.append("%.2E\t" %(del_23[int(j*(steps+2) + i*steps/printsteps)]))
        output.append("%.2E\t" %(del_12[int(j*(steps+2) + i*steps/printsteps)]))
        output.append("%.2E\t" %(del_32[int(j*(steps+2) + i*steps/printsteps)]))
        output.append("%.2E\t" %(del_34[int(j*(steps+2) + i*steps/printsteps)]))
        output.append("%.2E\t" %(del_43[int(j*(steps+2) + i*steps/printsteps)]))
        output.append("%.2E\t" %(del_4R[int(j*(steps+2) + i*steps/printsteps)]))
        output.append("%.2E\t" %(del_f[int(j*(steps+2) + i*steps/printsteps)]))
        output.append("%.2E\t" %(del_Pd[int(j*(steps+2) + i*steps/printsteps)]))
        output.append("%.2E\t" %(del_UR[int(j*(steps+2) + i*steps/printsteps)]))
        output.append("%.2E\t" %(del_R4[int(j*(steps+2) + i*steps/printsteps)]))
        output.append("%.2E\t" %(del_RU[int(j*(steps+2) + i*steps/printsteps)]))
        output.append("\n")


# write output to file
    FILE = open(filename,"w")          # open file in write-mode
    FILE.writelines(output)        # write all lines at once
    FILE.close()        # close file
```

A.1.5   Plot Results

```python
import pylab
from pylab import *
import matplotlib.pyplot as plt
import numpy as np
from numpy import *
from mpl_toolkits.mplot3d import Axes3D
import matplotlib.ticker as ticker

# import my own functions
from function_saturated_gaussian_simulation import saturated_gaussian_simulation

def plot_simulation(fig, printsteps, points, w, K_Pu, decay, N_G, N_L1, N_L2,
N_L3, N_L4, L):

#########################################

    if L == 1:
        N_0=N_L1[0]
    if L == 2:
        N_0=N_L2[0]
    if L == 3:
        N_0=N_L3[0]
    if L == 4:
        N_0=N_L4[0]

#############   subplots in one row   ###############

# left subplot
    if L == 1:
        ax = fig.add_subplot(421, projection="3d")
    if L == 2:
        ax = fig.add_subplot(423, projection="3d")
    if L == 3:
        ax = fig.add_subplot(425, projection="3d")
    if L == 4:
        ax = fig.add_subplot(427, projection="3d")

# color and appearance of plot background
    fig.patch.set_facecolor("w")        # colour of outer box
    ax.dist=50
# labels
    ax.set_xlabel(" \n\n\n Storage Time\n(s)", size=6)
    ax.set_ylabel(" \n\n\n\n Wavelength\n(nm)", size=6)
    ax.set_zlabel(" \n\n\n Number of\nProduct Ions", size=6)

# grid
```

```python
    ax.grid("off")

# tics
    # set tic positions manually
    ax.w_yaxis.set_major_locator(ticker.MaxNLocator(nbins=5))
    ax.w_zaxis.set_major_locator(ticker.MaxNLocator(nbins=5))
    # set format of tic labels
    ax.yaxis.set_major_formatter(FormatStrFormatter("%04.3f"))

# plot data from file
    x, y1, y2, z = np.loadtxt("level_%d.dat" %L, usecols=(0,11,13,3), unpack=1,
skiprows = 1)
    y = []
    j = len(y1)
    for i in range(j):
        y.append((y1[i]+y2[i])*1E9)

    ax.plot(x, y, z, marker = "o", linestyle = "None", markerfacecolor = "r",
markeredgecolor = "r", markersize = 0.5, label = "python simulation")

# plot function
    # get min and max values for x axis of plot above
    x_min, x_max = ax.get_xlim()
    # range for x values, notation:  (min, max, number_of_intervals)
    xs = np.linspace(x_min, x_max, 2*printsteps+1)
    y_min, y_max = ax.get_ylim()
    ys = np.linspace(y_min, y_max, 4*points+1)
    X, Y = np.meshgrid(xs, ys)
    z_min, z_max = ax.get_zlim()
    Z, saturation = saturated_gaussian_simulation(X, Y, y_min, y_max, N_G,
N_O, K_Pu, decay, w, z_max)

    wframe = ax.plot_wireframe(X, Y, Z, rstride=2, cstride=2, color = "g",
linewidth = 0.2, label = "saturated gaussian")

#########################################

# right subplot
    if L == 1:
        ax = fig.add_subplot(422)
    if L == 2:
        ax = fig.add_subplot(424)
    if L == 3:
        ax = fig.add_subplot(426)
    if L == 4:
        ax = fig.add_subplot(428)

# plot range
```

```python
    ax.set_ylim([0, N_G[0]])
# python will find up to nbins nice tic positions itself
    ax.xaxis.set_major_locator(ticker.MaxNLocator(nbins=5))
    ax.yaxis.set_major_locator(ticker.MaxNLocator(nbins=5))
# set format of tic labels
    ax.xaxis.set_major_formatter(FormatStrFormatter("%04.3f"))

    cset = ax.contourf(Y, Z, X)
    ax.set_xlabel("\nWavelength (nm)", size=6)
    ax.set_ylabel("Number of Product Ions", size=6)
    ax.text(y_min+0.001, N_G[0]*0.7, "Saturation:  %.4f\nN_L%d(0) = %d Ions\nMax
= %d Ions" %(saturation*x_max, L, N_0, z_max), size=6)
```

A.1.6 Saturated Gaussian

```python
import numpy as np

def saturated_gaussian_simulation(X, Y, y_min, y_max, N_G, N_0, K_Pu, decay,
w, z_max):

    ##########################################

    a=N_0/N_G[0]*K_Pu*(1-decay)*0.2
    a_x=a*X          # "deviation from gauss"
    yc=(y_min + y_max)/2        # position
    b=w*1e9        # standard deviation
    c=0        # baseline
    d=-N_G[0]        # line strength

    print ("Saturation:\n a = %.4f" %a)

    return d*(np.exp(-a_x*np.exp(-(Y-yc)**2/(2*b**2)))-1)+c, a        # saturated
gaussian, saturation
```

A.2 CH_5^+ Measurements

The N_0 values obtained from the fits for measurements at different laser powers are shown in Figure A.1. For the two measurements at 10.0 K there seems to be a trend of increasing line strength with higher laser power, although most data points can be considered constant within the large error bars. For the measurement at 24.8 K the data points seem to scatter around a constant value but they do not match within the errors obtained from the fit. Some care has to be taken when interpreting this figure for two reasons: The saturation of the observed lines was very low for the measurement series at 10.0 K, making the fit of the line strength not reliable; and the number of parent ions inserted in the trap was not constant, which is thought to have an effect on the line strength as well. The effect of the number of parent ions can be reduced by taking the ratio $\frac{\text{line strength}}{\text{background}}$ instead.

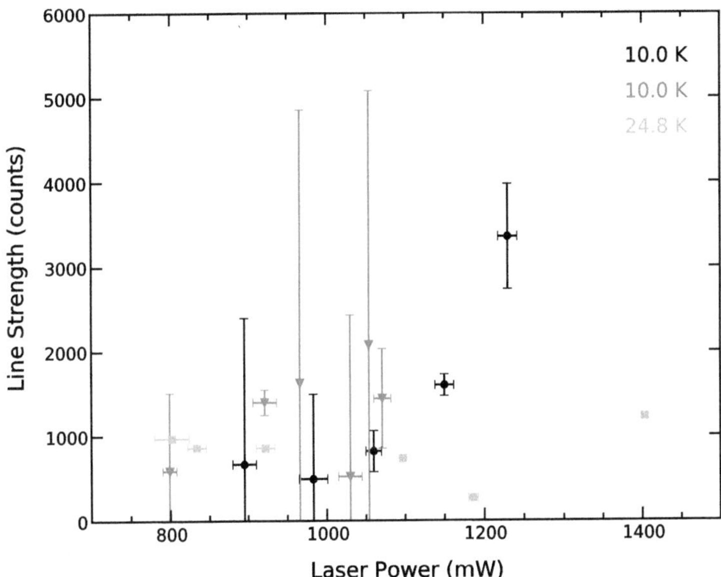

Figure A.1: Dependence of the line strength N_0 on the laser power. Three measurement series at two different nominal trap temperatures (10.0 and 24.8 K) are shown.

The saturation is expected to increase linearly with time. The enlarged view in Figure A.2 of the results shown in Figure 8.15 suggests this as well. However, the quality of the measurements is too low to derive ensured results.

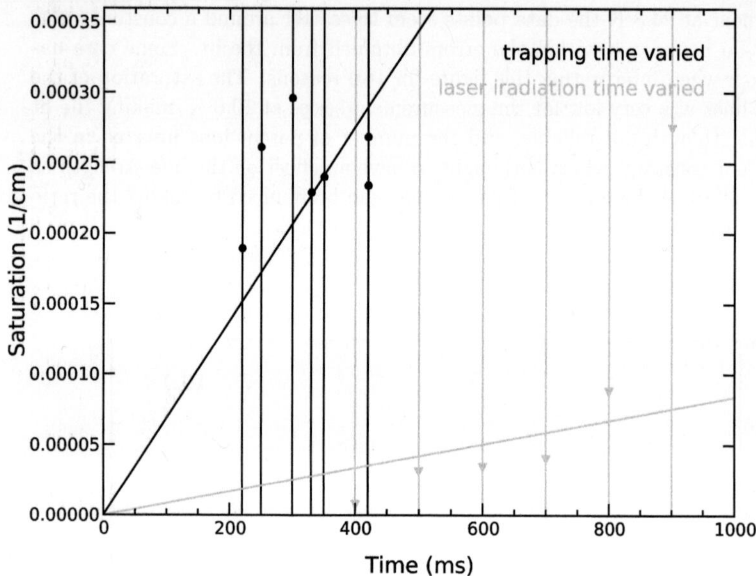

Figure A.2: Enlarged view of the dependence of the saturation a on the irradiation and the storage time. Two measurement series are shown, for one series the storage time of the ions was varied, for the other the laser irradiation time.

A.3 Construction of the Readout Electronics Housing for the new Daly Detectors

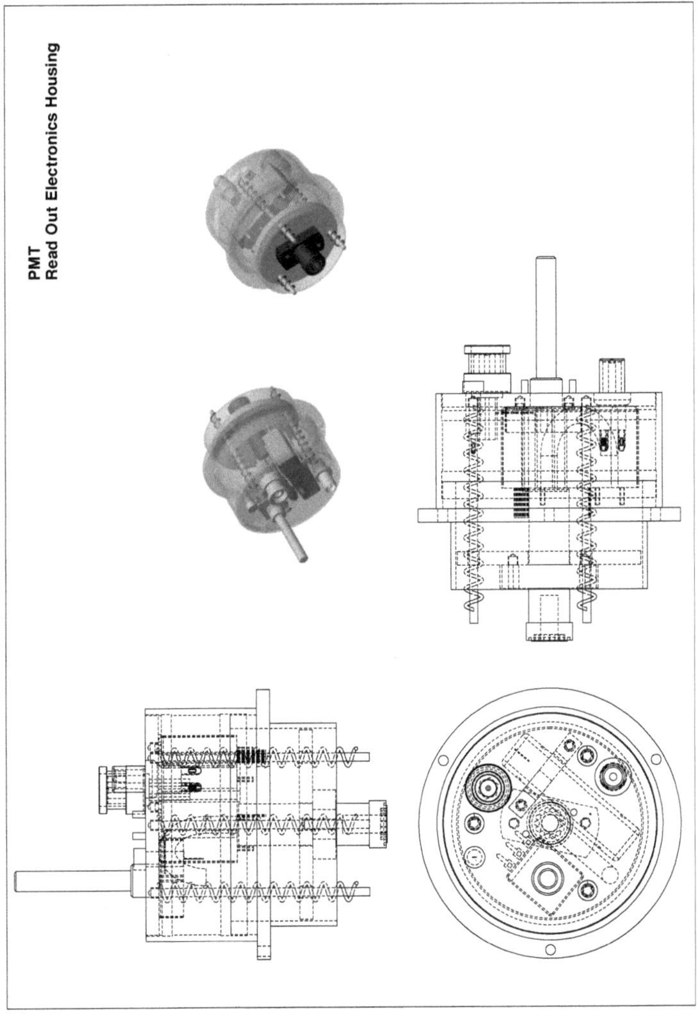

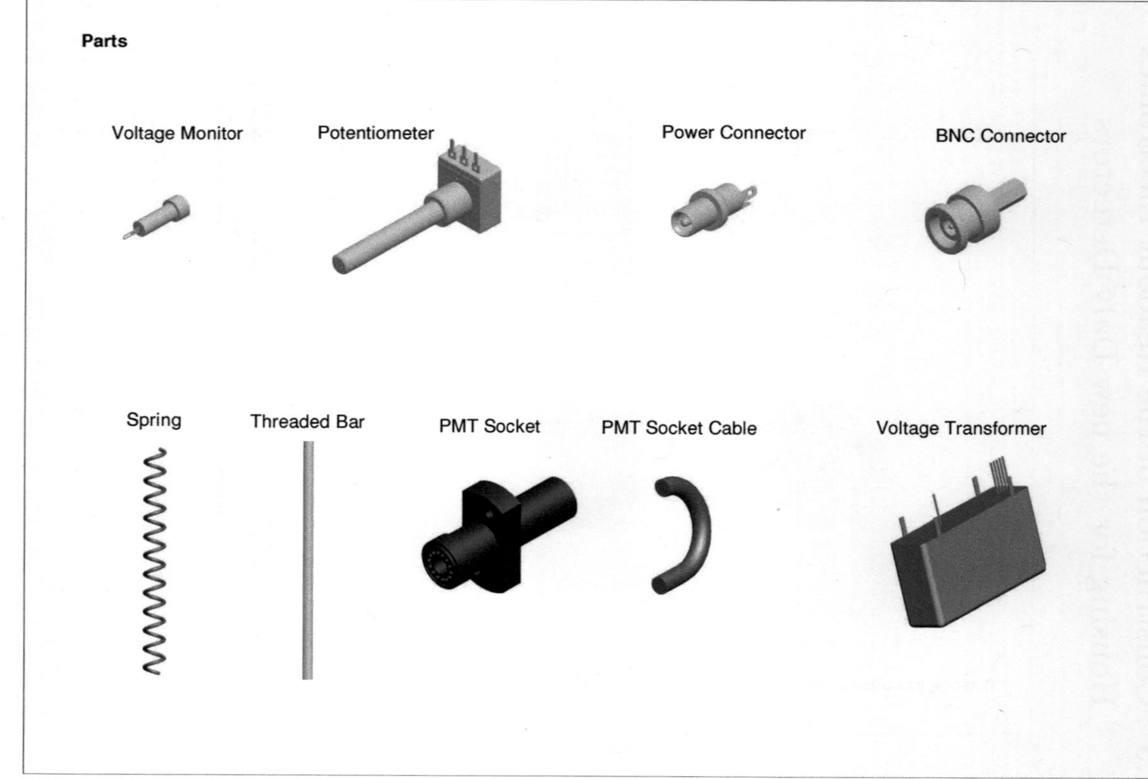
Parts
Voltage Monitor
Potentiometer
Power Connector
BNC Connector
Spring
Threaded Bar
PMT Socket
PMT Socket Cable
Voltage Transformer

Assembly

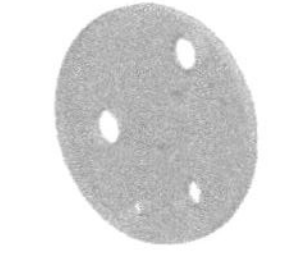

1. Back part with holes for the elctric connectors. The inner side is painted black.

2. Insert the connectors for BNC and power, the voltage monitor, and the potentiometer.

3. Fix the voltage transformer. Insert the threaded rods and the springs.

4. Build the electronic connections. (Two layers of shrinkable tubing over the connectors prevent light from entering.)
Fix the PMT socket to its base plate. Put the base plate on the threaded rods and secure it with nuts.

5. Srew the three parts of the housing together.

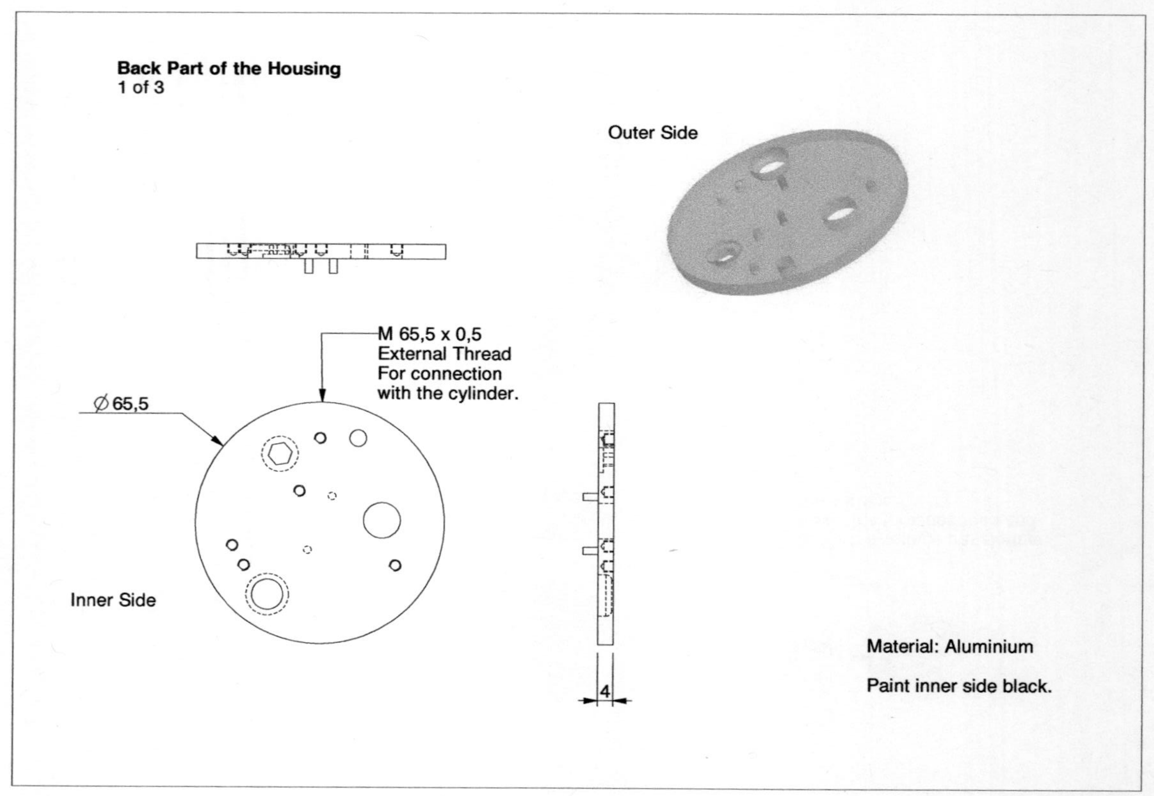
Back Part of the Housing
1 of 3
Outer Side
M 65,5 x 0,5
External Thread
For connection
with the cylinder.
Ø 65,5
Inner Side
Material: Aluminium
Paint inner side black.
4

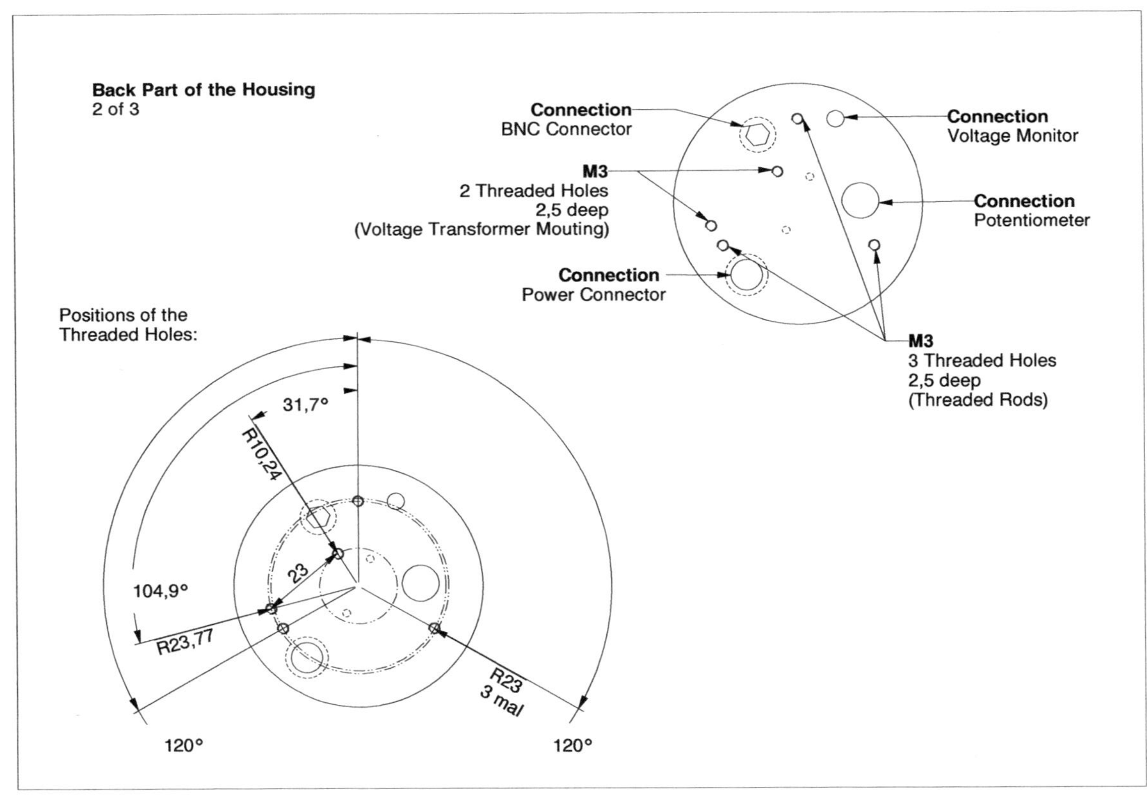

Back Part of the Housing
2 of 3
Connection
BNC Connector
Connection
Voltage Monitor
M3
2 Threaded Holes
2,5 deep
(Voltage Transformer Mouting)
Connection
Potentiometer
Connection
Power Connector
M3
3 Threaded Holes
2,5 deep
(Threaded Rods)
Positions of the
Threaded Holes:
31,7°
R10,24
104,9°
23
R23,77
120°
R23
3 mal
120°

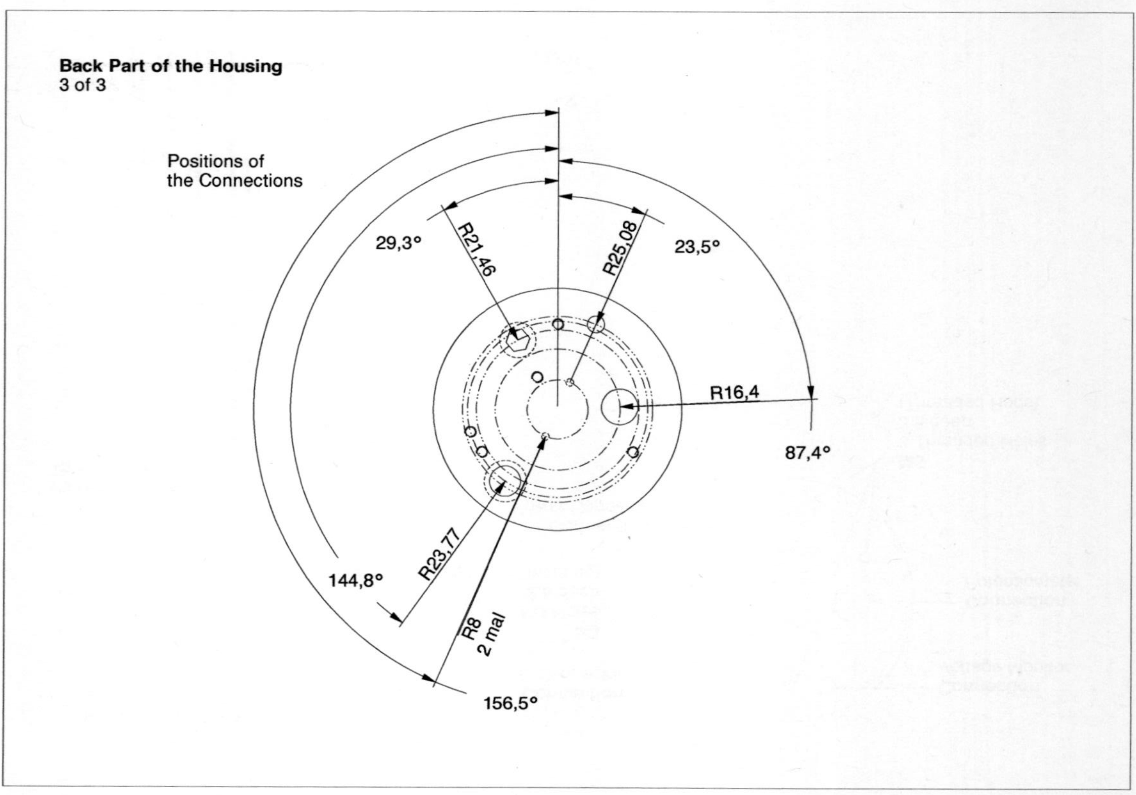

Back Part of the Housing
3 of 3
Positions of
the Connections
29,3°
R21,46
R25,08
23,5°
R16,4
87,4°
144,8°
R23,77
R8
2 mal
156,5°

Threaded Rod

M3

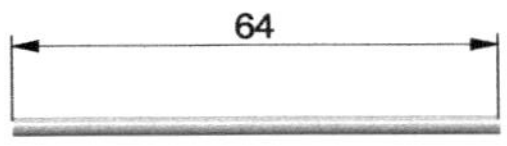

64

needed 3 times

Spring

Any spring that fits over an M3 rod may be used. The outer diameter should be as small as possible.

needed 3 times

Front part of the Housing

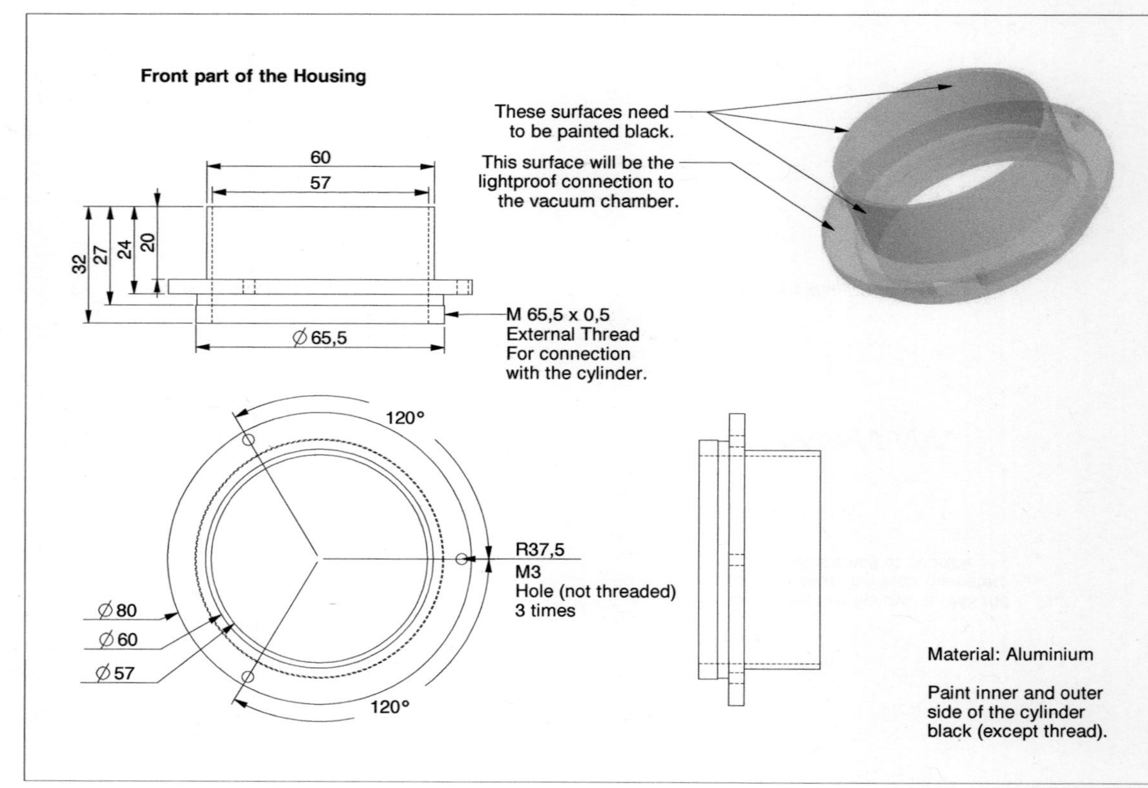

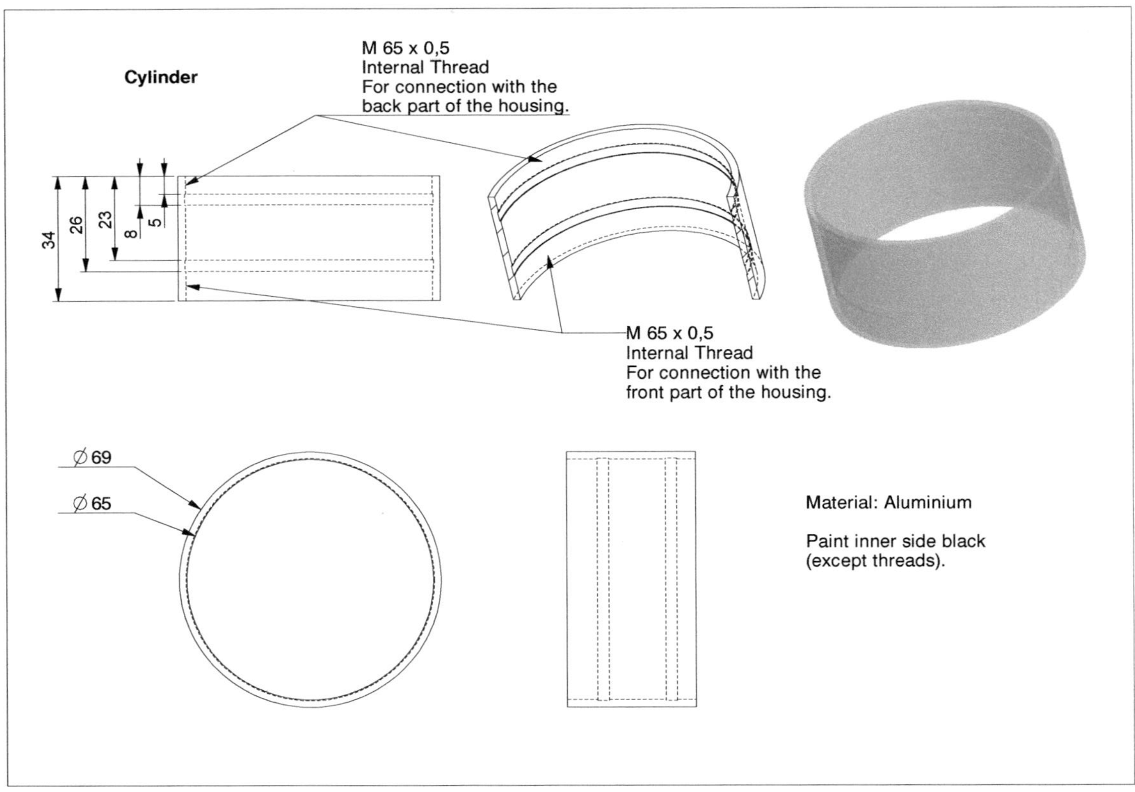

Cylinder
M 65 x 0,5
Internal Thread
For connection with the
back part of the housing.
M 65 x 0,5
Internal Thread
For connection with the
front part of the housing.
34
26
23
8
5
Ø 69
Ø 65
Material: Aluminium
Paint inner side black
(except threads).

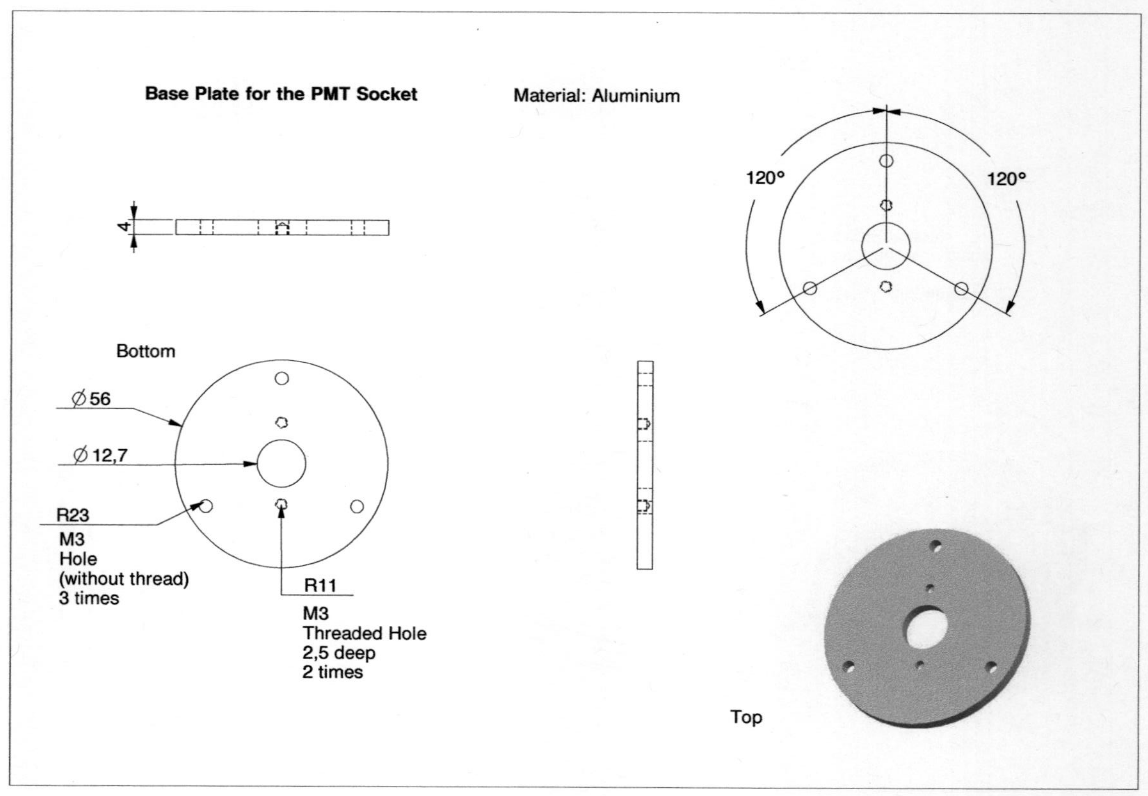

Base Plate for the PMT Socket
Material: Aluminium
120°
120°
4
Bottom
Ø 56
Ø 12,7
R23
M3
Hole
(without thread)
3 times
R11
M3
Threaded Hole
2,5 deep
2 times
Top

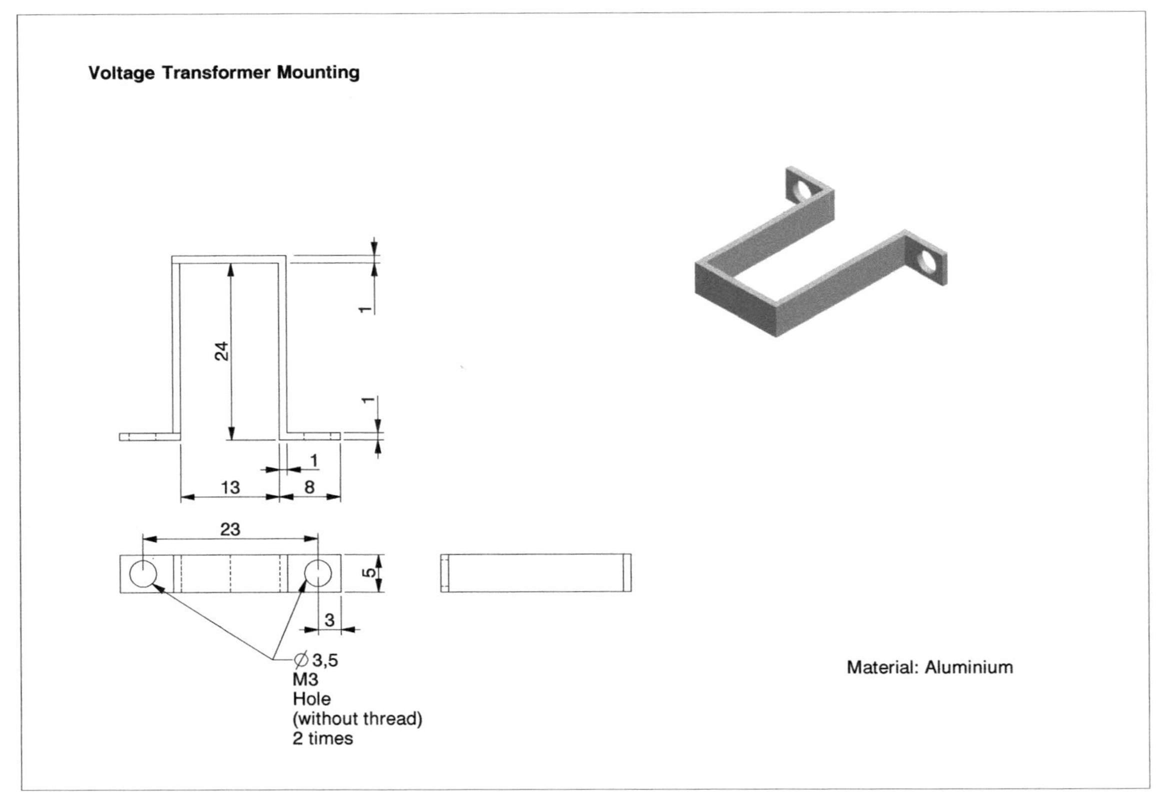

Voltage Transformer Mounting
24
1
1
13
8
1
23
5
3
∅ 3,5
M3
Hole
(without thread)
2 times
Material: Aluminium

A.4 Production of p-H$_2$

A.4.1 Continuous Flow

Table A.1 provides the valve settings in the para generator tube system for the production of p-H$_2$ in continuous flow mode. The constant pumping on the tubes (valve 8) ensures, that the p-H$_2$ is staying in the tubes only for short times. As iron oxide is a catalyst and the tubes and gas reservoirs are made of stainless steel, long stays in the tube system might lead to a back conversion. The converter box is kept at temperatures between 15 and 20 K for the whole process.

A.4.2 Freeze Out

Table A.2 shows the valve settings in the para generator tube system for the production of p-H$_2$ in freeze out mode. At the beginning of the production process all tubes and storage bottles are evacuated and the converter box is cooled down to $\approx$ 12 K. The freeze out of the H$_2$ into the converter box should happen slowly.

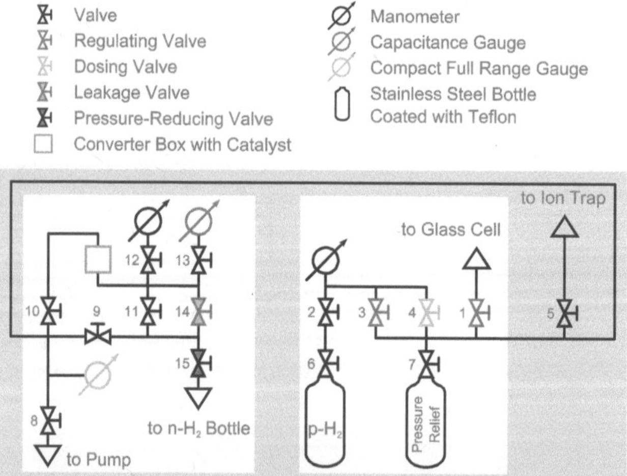

Figure A.3: Schematic of the para hydrogen generator tubes. The valves and connections shown on the left are needed for the continuous flow method. For production of p-H$_2$ by the freeze out method described in Section 5.1.2 the panel depicted on the right side was set up.

Table A.1: Settings of the valves in the para generator tube system in the different steps of continuous flow production of p-H_2: o = open, x = closed, a = adjust. For valve numbers, see Fig. A.3.

valve number	5	8	9	10	11	14	15	leakage valve to 22-pole trap chamber
initial settings	o	o	x	o	x	x	x	o
adjust H_2 flux		a				a	o	

Table A.2: Settings of the valves in the para generator tube system in the different steps of freeze out production of p-H_2: o = open, x = closed, a = adjust, os = open slowly. For valve numbers, see Fig. A.3.

valve number	2	3	4	5	6	8	9	10	11	14	15	converter temperature
initial settings	o	o	x	x	o	x	o	x	x	x	x	12 K
fill left storage bottle with 2 bar n-H_2											a	
freeze out n-H_2 from the tube system to the converter box		x						o	o		x	
freeze out n-H_2 from the storage bottle to the converter box			os									
pump out tubes (for one hour)		o	x			o		x	x			
pump out tube between valve 10 and converter box (for one minute)							x	o				
stop pumping						x						
slowly increase heating voltage for the cold head up to a value between 15 and 20 V until desired evaporation pressure is reached (250 to 750 mbar)												
close valve before manometer		x										17 to 21 K
close bottle				x								
switch off cold head and heating						o	o		o			warming up

A.5 Fluctuations in Raman Spectra

Figures A.4 and A.5 show an example of the intensity fluctuations within one Raman measurement. In Figure A.4 iterations one to five of ten were averaged, while Figure A.5 shows iterations six to ten. Especially in the o-H_2 line, the shifts of baseline and intensity over the measurement time of 30 minutes become evident. Their origin is not yet identified.

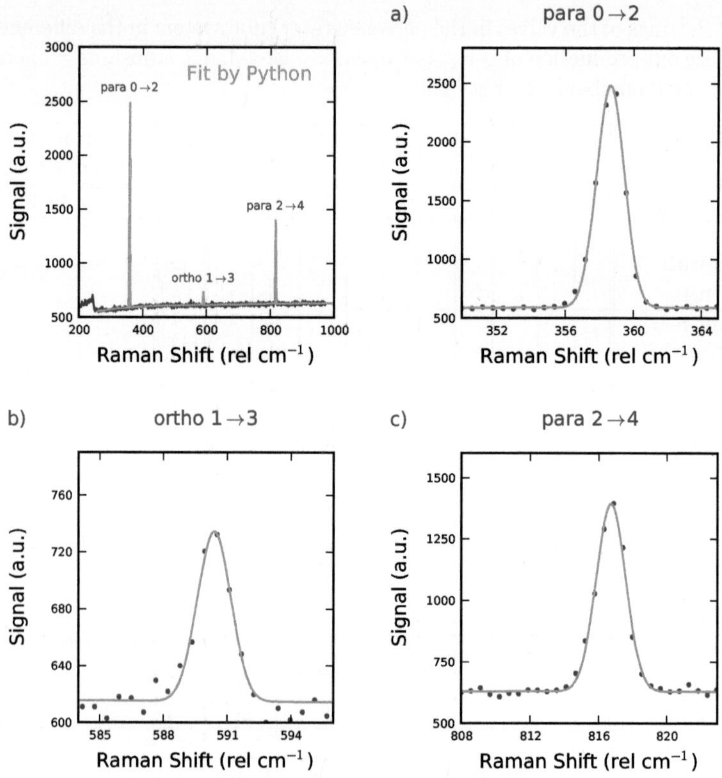

Figure A.4: Average over the first five iterations of the Raman measurement shown in Fig. 5.16.

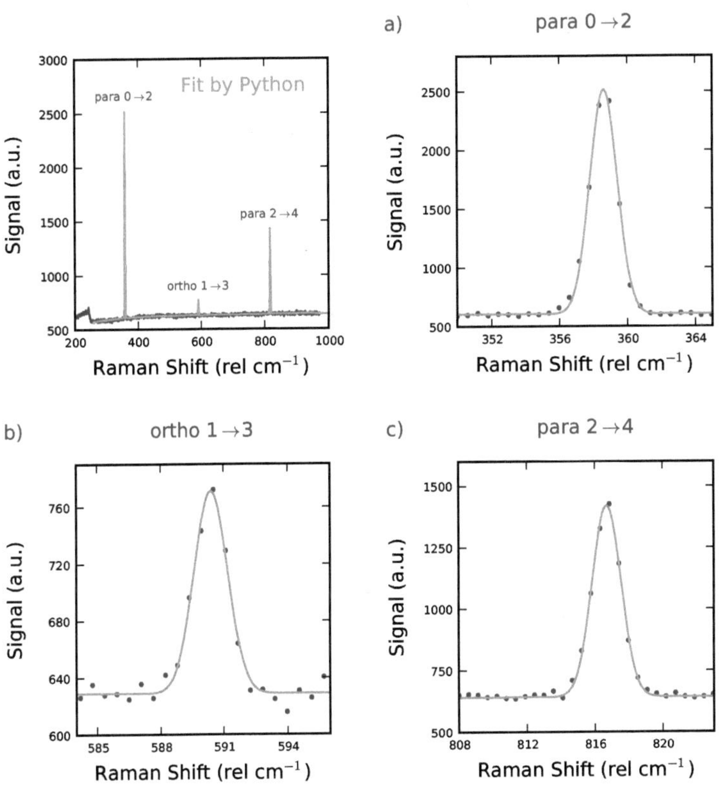

Figure A.5: Average over the last five iterations of the Raman measurement shown in Fig. 5.16.

A.6 $N^+ + H_2$

A.6.1 Reactions with HD

Figure 6.6 shows the lower detection limit for o-H_2 impurities, that is based on predictions and measurements by Dieter Gerlich [53]. When H_2 is admitted to the ion trap, it always contains some HD molecules. Thus, the following reactions can occur in experiments with $N^+ + H_2$:

$$N^+ + HD \;\rightleftharpoons\; NH^+ + D$$
$$\rightleftharpoons\; ND^+ + H$$

The low temperature behaviour of the ratecoefficient for the destruction of N^+ by HD as observed by Gerlich can be described by the function

$$k(T) \;=\; 3.5 \times 10^{-10} e^{-\frac{11\,\mathrm{K}}{T}} \; \frac{\mathrm{cm}^3}{\mathrm{s}} \;. \tag{A.1}$$

For temperatures higher than 30 K, the results deviate from this curve. These deviations are comparable to the deviations observed for o-H_2 and p-H_2. They

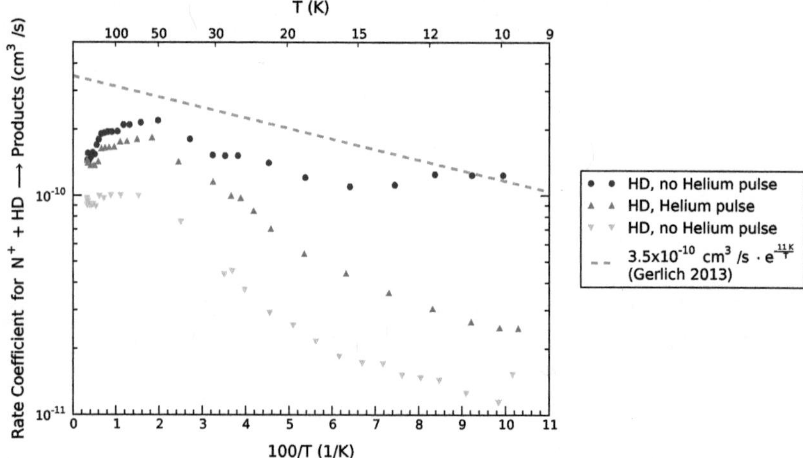

Figure A.6: Temperature dependence of the rate coefficient for the reaction of N^+ with HD. The *symbols* show measurements performed recently in Cologne by Sven Fanghänel. Three different measurement series with and without helium buffer gas pulse were performed. The origin of the large deviations is unclear so far. All measurement results lie below the function given by Gerlich [53] *(dashed line)*, which is therefore given as lower detection limit in Fig. 6.6.

are probably all caused by the fine-structure states of N^+. Figure 6.6 compares function (A.1) to first results from the cologne laboratories (by Sven Fanghänel). The Cologne measurements on N^+ + HD show even worse problems with the reproducibility, than the experiments on N^+ + H_2. As all of them lie below the curve given by Gerlich, the latter is shown as lower detection limit in Figure 6.6.

A.6.2   Effects of the Helium Buffer Gas

Figures A.7 to A.15 show comparisons of N^+ time evolutions for the different measurement series at different nominal trap temperatures, reaching from 10 to 267 K. For better comparability of the measurements, the time axis of each curve was corrected for the change in the number density of n-H_2, [n-H_2], and the N^+ axis for the change of the number of ions at $t = 0$, N_0. As could already be seen from Figure 6.10, no trend for the reaction speed is visible for the individual temperatures (the influence of the temperature itself shows nicely). Neither is an influence of the He density on the bending of the curves observable.

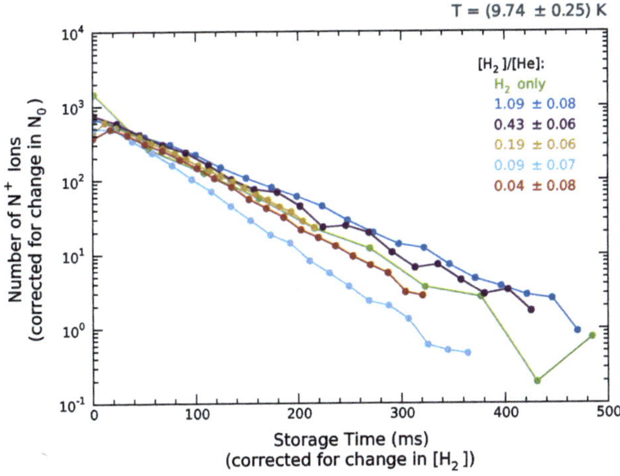

Figure A.7: Decrease of the number of N^+ ions over the storage time in reaction with n-H_2 for different amounts of helium buffer gas in the trap at a nominal trap temperature of 9.7 K.

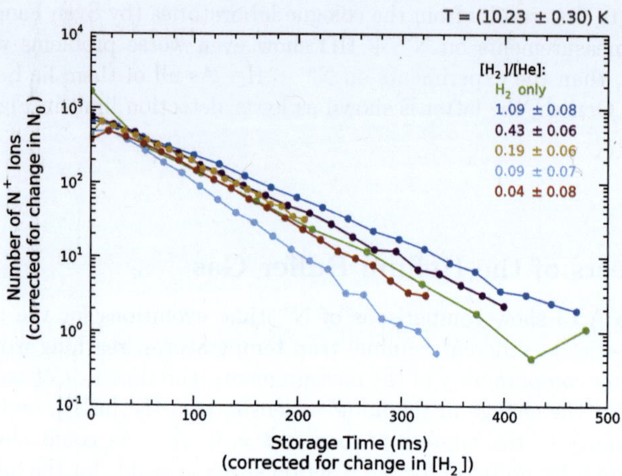

Figure A.8: Decrease of the number of N^+ ions over the storage time in reaction with n-H_2 for different amounts of helium buffer gas in the trap at a nominal trap temperature of 10.2 K.

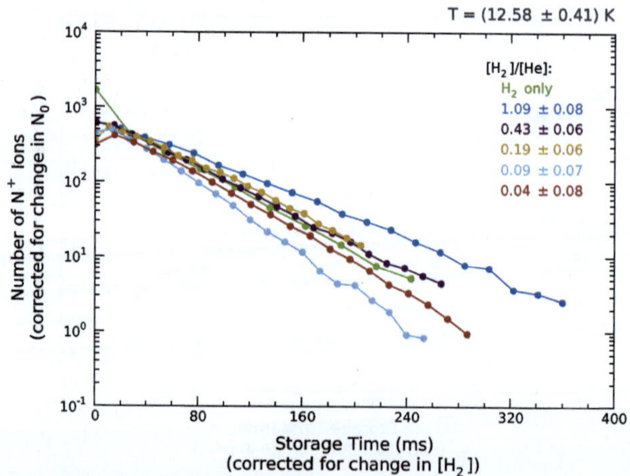

Figure A.9: Decrease of the number of N^+ ions over the storage time in reaction with n-H_2 for different amounts of helium buffer gas in the trap at a nominal trap temperature of 12.6 K.

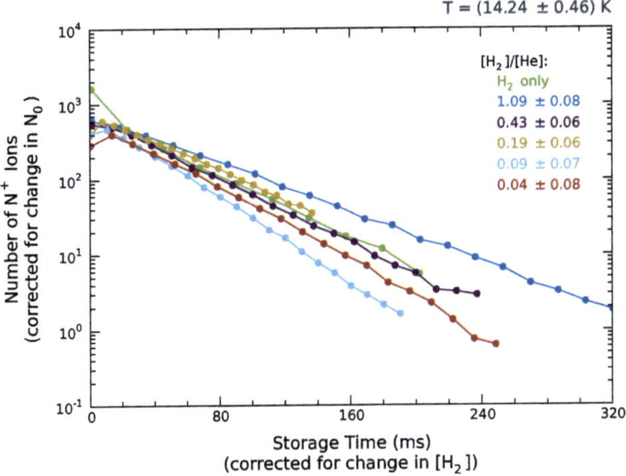

Figure A.10: Decrease of the number of N$^+$ ions over the storage time in reaction with n-H$_2$ for different amounts of helium buffer gas in the trap at a nominal trap temperature of 14.2 K.

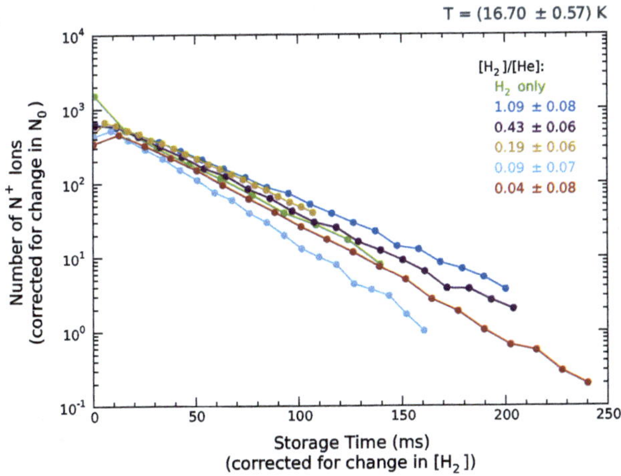

Figure A.11: Decrease of the number of N$^+$ ions over the storage time in reaction with n-H$_2$ for different amounts of helium buffer gas in the trap at a nominal trap temperature of 16.7 K.

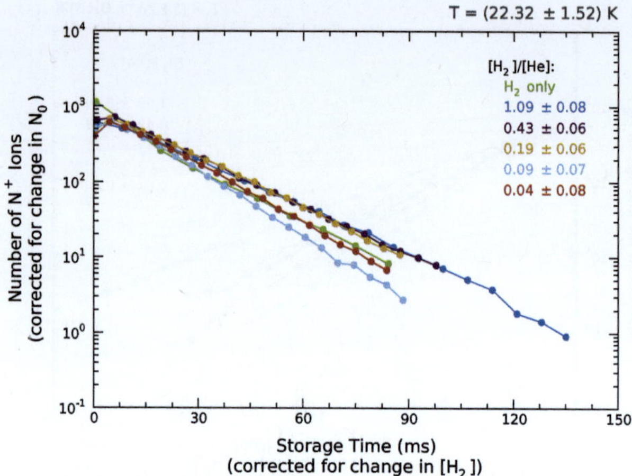

Figure A.12: Decrease of the number of N^+ ions over the storage time in reaction with n-H_2 for different amounts of helium buffer gas in the trap at a nominal trap temperature of 22 K.

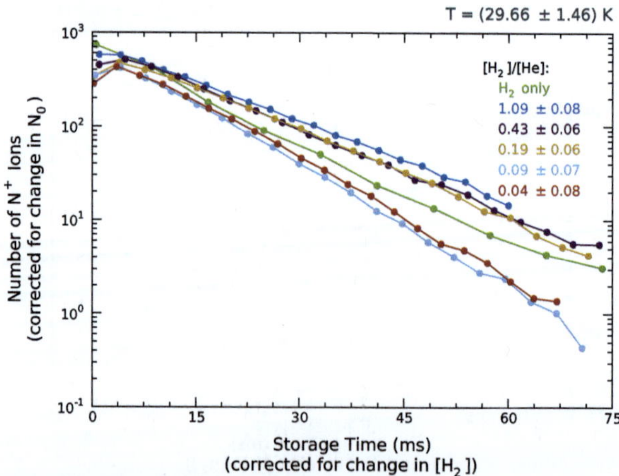

Figure A.13: Decrease of the number of N^+ ions over the storage time in reaction with n-H_2 for different amounts of helium buffer gas in the trap at a nominal trap temperature of 30 K.

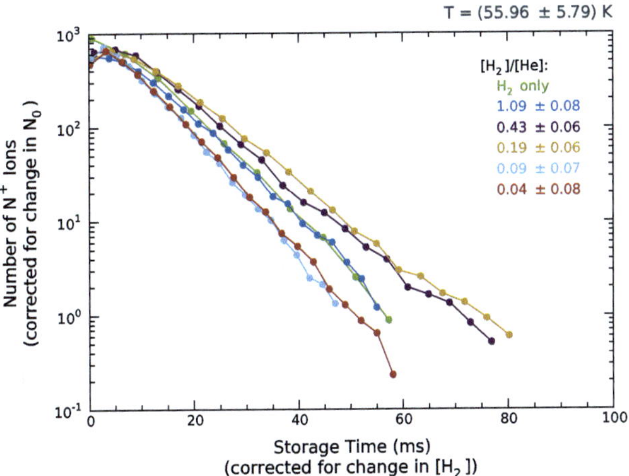

Figure A.14: Decrease of the number of N^+ ions over the storage time in reaction with n-H_2 for different amounts of helium buffer gas in the trap at a nominal trap temperature of 56 K.

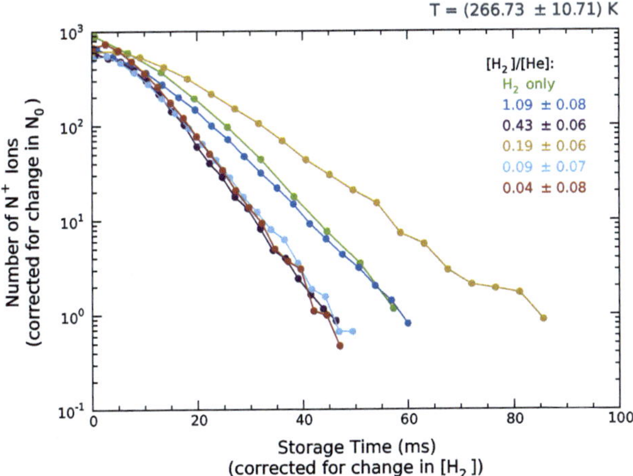

Figure A.15: Decrease of the number of N^+ ions over the storage time in reaction with n-H_2 for different amounts of helium buffer gas in the trap at a nominal trap temperature of 267 K.

A.6.3　Calibration of the RF Amplitude

The radio frequency applied to the rods of the 22-pole trap is produced by an RF generator. From the generator, the signal is fed into the vacuum chamber via a vacuum feedthrough. Inside the vacuum chamber, a coil is connected to the feedthrough. Inside of this coil, another coil is located, that picks up the RF signal and feeds it to the 22-pole. This setup prevents thermal contact between the 22-pole rods that have to be at trap temperature and the room temperature feedthrough. Inside the second coil, an antenna is located, which is connected to another vacuum feedthrough. Here, the RF amplitude can be probed, when the vacuum chamber is closed.

Within this work, calibration measurements were performed on the RF amplitude. In the first step, the correlation between the potentiometer setting on the RF generator and the actual amplitude on the inner coil was measured. For this purpose, the vacuum chamber had to be open and a probe was connected to the coil. Both aspects influence the RF amplitudes. To correct for the influence of the probe, the signal picked up by the antenna was measured with and without the probe connected to the coil. To correct for the influence of the open vacuum chamber, the signal at the vacuum feedthrough of the antenna was measured with open and with closed vacuum chamber. The calibration measurements are presented in Figure A.16. The following functions were fitted to the data (the values of the fit parameters are provided in Figure A.16):

$$RF_{\text{coil}}^{\text{chamber open}} = a_{1,a} \cdot RF_{\text{dial}} + b_{1,a} \ , RF_{\text{dial}} < 2.3$$

$$RF_{\text{coil}}^{\text{chamber open}} = a_{1,b} \cdot RF_{\text{dial}} + b_{1,b} \ , RF_{\text{dial}} > 2.4$$

$$RF_{\text{antenna}}^{\text{chamber open}} = a_2 \cdot RF_{\text{antenna}}^{\text{chamber open, probe connected to coil}} + b_2$$

$$RF_{\text{feedthrough}}^{\text{chamber closed}} = a_3 \cdot RF_{\text{feedthrough}}^{\text{chamber open}} + b_3$$

This leads to:

$$RF_{\text{coil}}^{\text{chamber closed, probe not connected to coil}} = a_{1,b} a_2 a_3 \cdot RF_{\text{dial}} + a_3 a_2 b_{1,a} + a_3 b_2 + b_3$$

$$RF_{\text{coil}}^{\text{chamber closed, probe not connected to coil}} = (205.45 \pm 3.01)\,\text{V} \cdot RF_{\text{dial}} - (471.15 \pm 7.29)\,\text{V}$$

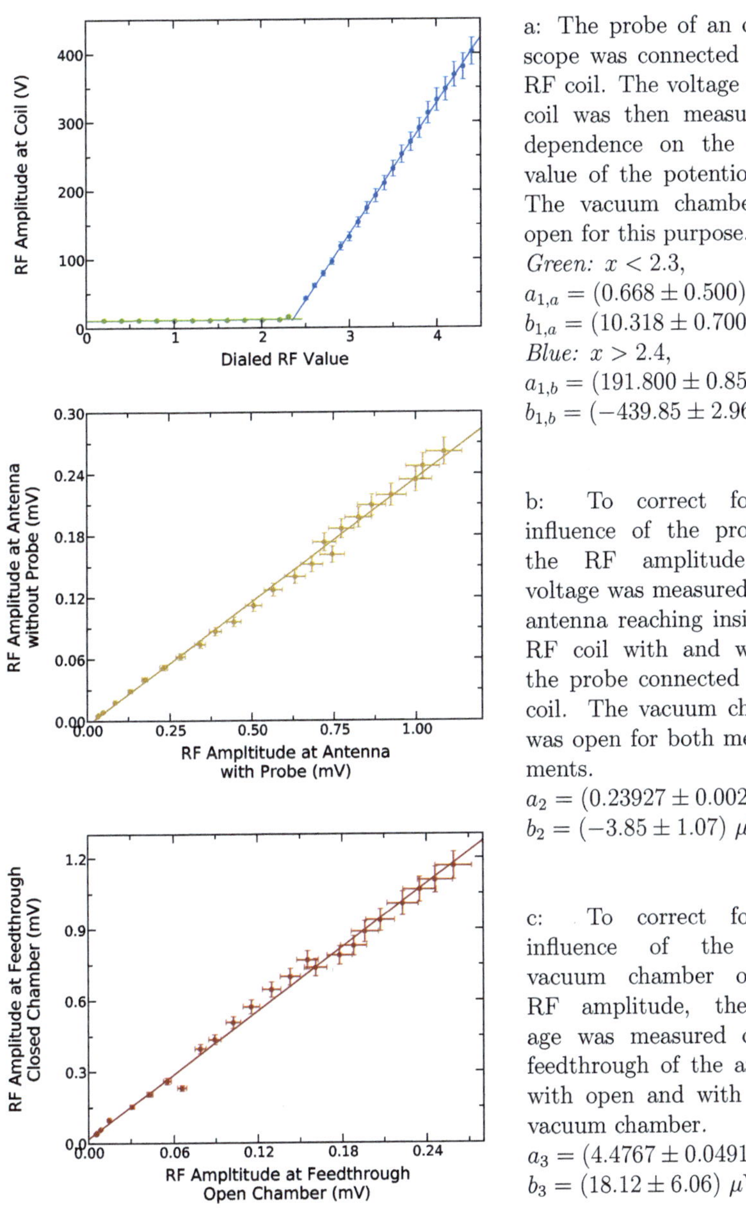

a: The probe of an oscilloscope was connected to the RF coil. The voltage on the coil was then measured in dependence on the dialed value of the potentiometer. The vacuum chamber was open for this purpose.
Green: $x < 2.3$,
$a_{1,a} = (0.668 \pm 0.500)$ V
$b_{1,a} = (10.318 \pm 0.700)$ V
Blue: $x > 2.4$,
$a_{1,b} = (191.800 \pm 0.858)$ V
$b_{1,b} = (-439.85 \pm 2.96)$ V

b: To correct for the influence of the probe on the RF amplitude, the voltage was measured on an antenna reaching inside the RF coil with and without the probe connected to the coil. The vacuum chamber was open for both measurements.
$a_2 = (0.23927 \pm 0.00205)$
$b_2 = (-3.85 \pm 1.07)$ μV

c: To correct for the influence of the open vacuum chamber on the RF amplitude, the voltage was measured on the feedthrough of the antenna with open and with closed vacuum chamber.
$a_3 = (4.4767 \pm 0.0491)$
$b_3 = (18.12 \pm 6.06)$ μV

Figure A.16: RF calibration. Fit function: $f(x) = a \cdot x + b$. The jumps in the curve are artifacts from the oscilloscope. They happen, when the device switches between two measuring ranges.

Danksagung

Ich möchte mich an dieser Stelle ganz herzlich bei allen Menschen bedanken, die mich bei meiner Doktorarbeit unterstützt haben.

Allen voran bei Prof. Dr. Stephan Schlemmer, für die Betreuung meiner Arbeit und die immer neuen, faszinierenden Fragestellungen; für angeregte Diskussionen, lehrreiche Bastelstunden im Labor (auch schon mal im Karnevalskostüm) und die fantastische Arbeitsatmosphäre.
Vielen Dank an Prof. Dr. Andreas Wolf für die Übernahme des Zweitgutachtens für meine Arbeit.
Bei Oskar Asvany, der mir bei meiner Arbeit immer mit Rat und Tat zur Seite stand, als unerschöpfliche Quelle des Wissens. Danke für die ansteckende gute Laune, für das Korrekturlesen meiner Arbeit und natürlich für die Mitwirkung im Prüfungskomitee als Berichterstatter.
Vielen Dank an Prof. Dr. Joachim Krug für den Vorsitz im Prüfungskomitee.

Der Dank für die großartige Arbeitsatmosphäre gebührt natürlich der gesamten Arbeitsgruppe Laborastrophysik der Uni Köln, ihr seid einmalig.
Ganz besonders möchte ich mich bei Volker Lutter bedanken, für das gemeinsam durchgestandene Studium, für klärende Diskussionen, wenn ich mich bei einem Problem festgelaufen hatte, für herrlich schlechte Scherze, für das geduldige Lesen der ersten unausgegorenen Schreibereien und natürlich dafür, dass er immer genau dann Schokolade suchen kam, wenn ich gerade eine Pause brauchte.
Vielen Dank auch an Monika Koerber, die auf jede Spektroskopiefrage eine Antwort und notfalls immer einen Bernath hat, unter Anderem für die gemeinsame Motivation zum Sport und das unermüdliche Lesen diverser Kapitel dieser Arbeit. An Lars Kluge, der immer -mal eben- Zeit hat, wenn ich einen Diskussionspartner brauche; natürlich auch für die vielen korrigierten Kapitel meiner Arbeit. Danke an alle derzeitigen und ehemaligen Büronachbarn für die immer gute Atmosphäre. Danke an Silvia Spezzano u. A. für die Versorgung mit italienischen Leckereien und an Doris Herberth für den nicht zu bremsenden Optimismus.
Vielen Dank auch an alle Laborkollegen. An Sandra Brünken für gute Antworten auf blöde Fragen, an Juliane Gerke und Mareke Sommer für die Einführungen zum Ramanspektrometer, an Henning Adams für die technische Unterstützung bei Umbaumaßnahmen, an Sven Fanghänel für die Fortführung der N^+ Messungen und natürlich an Alexander Stoffels für Dexter und Abendbrot an langen Messabenden.

Dank gebührt auch dem gesamten Team der Bonn-Cologne-Graduate-School of Physics and Astronomy für ihre Unterstützung. Allen voran Petra Neubauer-Guenther, die für alle Fragen ein offenes Ohr und für alle Probleme eine Lösung hat und mit der man sich wunderbar nach dem Essen verquatschen kann. Vielen Dank auch an meinen Bonner Mentor Prof. Dr. Dieter Meschede.

Ein ganz herzliches Dankeschön auch an alle heutigen und ehemaligen Mitglieder der Fachschaft Physik. Nicht nur für gemeinsame FS-Arbeit, sondern auch für Lauftreffs, Spieleabende und vieles mehr.

Vielen dank auch an die Arbeitsgruppe von Prof. Wolf am MPI für Kernphysik in Heidelberg, vor allem an Max Berg, Florian Grussie und Holger Kreckel für die Zeit, die ich mit Ihnen im Heidelberger Labor verbringen durfte.
Many thanks also to Prof. Dr. Juraj Glosík and his group at the Charles University in Prague, for the time I could spend in their labs. Special thanks to Illja Zymak and Michal Hejduk for the time we spent together at the experiments and the valueable discussions. Thanks to Petr Dohnal for the help with travel and accomodation, and to Peter Rubovič for the nice time we spent sightseeing.

Ein großes Dankschön auch an meine Familie. Allen voran an meinen Bruder Etienne für das unfassbar gewissenhafte Lesen meiner gesamten Arbeit, für die unfehlbare Spürnase für Normierungsfaktoren und den Blick eines Außenstehenden auf meine Probleme. Danke an meine Schwester Vanessa, an Etienne und meine Schwägerin Birthe für gemeinsames Renovieren, Nähen, Spielen und vieles mehr. Danke an meine Eltern und an Gerd und Waltraud für die Unterstützung, die Versorgung mit guten Büchern zum Abschalten, für's gemeinsame Shoppen, Quatschen und für Vollverpflegung, wann immer man bei ihnen einfällt. Danke an meine Omas Emmi und Else für die Versorgung mit allem Möglichen, von warmen Socken bis zu gefüllten Paprika, und an Anita und Manfred u. A. für abwechslungsreiche Rezepte.

Ich möchte mich auch bei allen Freunden und Bekannten bedanken, mit denen ich viele und doch zu wenige nette Stunden verbracht habe. Ganz besonders bei Stephanie Orbe für die gemeinsamen Nähaktionen.

Last but not least, der größte Dank gilt natürlich meinem Freund Jürgen Gogolin für die letzten zehn Jahre. Danke, für alles!

Ich versichere, dass ich die von mir vorgelegte Dissertation selbständig angefertigt, die benutzten Quellen und Hilfsmittel vollständig angegeben und die Stellen der Arbeit - einschließlich Tabellen, Karten und Abbildungen -, die anderen Werken im Wortlaut oder dem Sinn nach entnommen sind, in jedem Einzelfall als Entlehnung kenntlich gemacht habe; dass diese Dissertation noch keiner anderen Fakultät oder Universität zur Prüfung vorgelegen hat; dass sie - abgesehen von unten angegebenen Teilpublikationen - noch nicht veröffentlicht worden ist, sowie, dass ich eine solche Veröffentlichung vor Abschluss des Promotionsverfahrens nicht vornehmen werde. Die Bestimmungen der Promotionsordnung sind mir bekannt. Die von mir vorgelegte Dissertation ist von Herrn Professor Dr. Stephan Schlemmer betreut worden.

Köln, im Oktober 2013

Liste der Veröffentlichungen

1. Sabrina Gärtner, Jürgen Krieg, Andre Klemann, Oskar Asvany, and Stephan Schlemmer. Rotational Transitions of CH_2D^+ determined by high-resolution ir spectroscopy. *Astron. Astrophys.*, 516:L3, 2010.

2. Lars Kluge, Sabrina Gärtner, Sandra Brünken, Oskar Asvany, Dieter Gerlich, and Stephan Schlemmer. Transfer of a proton between H_2 and O_2. *Phil. Trans. R. Soc. A*, 370(1978):5041–5054, 2012.

3. F. Grussie, M. H. Berg, K. N. Crabtree, S. Gärtner, B. J. McCall, S. Schlemmer, A. Wolf, and H. Kreckel. The Low-temperature Nuclear Spin Equilibrium of H_3^+ in Collisions with H_2. *Astrophys. J.*, 759:21, 2012.

4. Sabrina Gärtner, Jürgen Krieg, Andre Klemann, Oskar Asvany, Sandra Brünken, and Stephan Schlemmer. High-Resolution Spectroscopy of CH_2D^+ in a Cold 22-Pole Trap. *J. Phys. Chem. A*, in press:10.1021/jp400258e.

Lebenslauf

Persönliche Daten:

Name:	Sabrina Gärtner
Adresse:	Meschenicher Str. 447
	50997 Köln
Geburtstag:	07.11.1981
Geburtsort:	Leverkusen-Opladen
Familienstand:	ledig
Eltern:	Doris Gärtner geb. Bros (04.11.1955)
	Rechtsanwalts-Fachangestellte
	Holger Gärtner (22.08.1951)
	Diplom-Kaufmann
Geschwister:	Etienne Gärtner (09.06.1983)
	Diplom-Physiker
	Vanessa Gärtner (25.12.1984)
	Kauffrau für Bürokommunikation

Schul- und Hochschulbildung:

1988 bis 1992	Grundschule Sandberg in Monheim
1992 bis 2001	Otto-Hahn-Gymnasium in Monheim
Mai 2001	Abitur: Note 1,4
	Mir wurde damit zusammenhängend die einjährige kostenlose Mitgliedschaft in der Deutschen Physikalischen Gesellschaft als Anerkennung für sehr gute Leistungen im Fach Physik verliehen.
ab WS 2001	Physikstudium an der Universität zu Köln
Januar 2005	Vordiplom: Note „gut"
November 2007	Erhalt des Diplom-Zwischenzeugnisses: Note „sehr gut"
Januar 2008	Beginn der Diplomarbeit zum Thema „Charakterisierung einer Ionen-Wolke in einem kalten 22-Pol Ionenspeicher"
Oktober 2008	Aufnahme in die „Bonn Cologne Graduate School of Physics and Astronomy"
Februar 2009	Diplom: Note „sehr gut"
März 2009	Beginn der Promotion

Bisherige Tätigkeiten:

Siemens-Rexroth: (Lohr am Main)

03.07.00 bis 21.07.00	Bürotätigkeit im Archiv für technische Zeichnungen

Bayer: (Leverkusen)

09.07.01 bis 31.08.01	Bürotätigkeit in der Abteilung Qualitätssicherung als Werksstudent
22.07.02 bis 13.09.02	Qualitätssicherung in der Produktion als Werksstudent

Universität zu Köln:

01.10.05 bis 31.12.05	Tutorium zur Vorlesung „Physik 1"
Dezember 2005	Schnupperuniversität Physik: Koordination der studentischen Hilfskräfte und Betreuung der Schülerinnen
08.05.06 bis 07.08.06	Tutorium zum Anfängerpraktikum und zur Vorlesung „Physik für Mediziner"
30.10.06 bis 17.02.07	Tutorium zu den Vorlesungen „Mathematische Methoden der Physik" und „Experimentalphysik 1"
08. und 09.02.07	Schnupperuniversität Physik: Koordination der studentischen Hilfskräfte und Betreuung der Schülerinnen
02.04.07 bis 13.07.07	Tutorium zur Vorlesung „Klassische Theoretische Physik 1"
19. und 20.10.07	Schnupperuniversität Physik: Koordination der studentischen Hilfskräfte und Betreuung der Schülerinnen
01.01.08 bis 30.06.09	Betreuung des Anfängerpraktikums im 1. Physikalischen Institut
25. und 26.09.08	Schnupperuniversität Physik: Organisation der Schülerversuche, sowie Durchführung und Betreuung eines Schülerversuchs
seit 01.07.09	Praktikumskoodination im 1. Physikalischen Institut
08. und 09.10.09	Schnupperuniversität Physik: Organisation der Schülerversuche, sowie Durchführung und Betreuung eines Schülerversuchs
07. und 08.10.10	Schnupperuniversität Physik: Organisation der Schülerversuche, sowie Durchführung und Betreuung eines Schülerversuchs
07. und 08.03.12	Schnupperuniversität Physik: Organisation der Schülerversuche, sowie Durchführung und Betreuung eines Schülerversuchs

Köln, Oktober 2013